T0309591

PROBLEMS AND SOLUTIONS

Nonlinear Dynamics, Chaos and Fractals

PROBLEMS AND SOLUTIONS

Nonlinear Dynamics, Chaos and Fractals

Willi-Hans Steeb

University of Johannesburg, South Africa

NEW JERSEY • LONDON • SINGAPORE • BEIJING • SHANGHAI • HONG KONG • TAIPEI • CHENNAI • TOKYO

Published by

World Scientific Publishing Co. Pte. Ltd.
5 Toh Tuck Link, Singapore 596224
USA office: 27 Warren Street, Suite 401-402, Hackensack, NJ 07601
UK office: 57 Shelton Street, Covent Garden, London WC2H 9HE

Library of Congress Cataloging-in-Publication Data
Names: Steeb, W.-H., author.
Title: Problems and solutions : nonlinear dynamics, chaos and fractals / Willi-Hans Steeb, University of Johannesburg, South Africa.
Description: Singapore ; Hackensack, NJ : World Scientific Publishing Co. Pte. Ltd., [2016] | 2016 | Includes bibliographical references and index.
Identifiers: LCCN 2016002804| ISBN 9789813109926 (hardcover ; alk. paper) | ISBN 9813109920 (hardcover ; alk. paper) | ISBN 9789813140875 (pbk ; alk. paper) | ISBN 9813140879 (pbk ; alk. paper) |
Subjects: LCSH: Nonlinear theories--Problems, exercises, etc. | Dynamics--Problems, exercises, etc. Chaotic behavior in systems--Problems, exercises, etc. | Fractals--Problems, exercises, etc.
Classification: LCC QA427 .S74 2016 | DDC 516.3/5--dc23
LC record available at http://lccn.loc.gov/2016002804

British Library Cataloguing-in-Publication Data
A catalogue record for this book is available from the British Library.

Printed in Singapore

Preface

The purpose of this book is to supply a collection of problems for nonlinear dynamics, chaos theory and fractals. Besides the solved problems supplementary problems are added. Each chapter contains an introduction with the essential definitions and explanations to tackle the problems. If necessary, other concepts are explained directly with the problem. Thus the material in this book is self-contained. The topics range in difficulty from elementary to advanced. Students can learn important principles and strategies required for problem solving. Lecturers will also find this text useful, either as a supplement or text, since concepts and techniques are developed in the problems.

In chapter 1, one-dimensional maps are introduced. Fixed points, periodic points and their stability play the central role in this chapter as well as the linearized equation. Exercises for chaotic maps such as the logistic map, tent map, Bernoulli map are provided. The Liapunov exponent to characterize chaos is introduced. Problems for linearization of nonlinear maps as well as invariants are provided. Exercises for the Newton map are also introduced. A number of problems deal with ergodic theory and the Frobenius-Perron integral equation.

In chapter 2, two-dimensional and higher dimensional maps are covered as well as complex maps. Exercises for both invertible and noninvertible maps are provided. The Newton map is also covered. Applications to differential forms are introduced. A section deals with boolean maps.

Chapter 3 is devoted to fractals, where fractal dimensions play a central role. For fractals, iterated functions systems are at the core of the construction. The Kronecker product is also introduced for the construction of fractals. Exercises cover the Cantor set, Sierpinski triangle, Sierpinski carpet, Koch curve, Menger sponge, etc.. A number of problems deal with the Mandelbrot set and Julia set.

For some solutions of the problems software such as C++, SymbolicC++, Maxima and R are utilized.

The International School for Scientific Computing (ISSC) provides certificate courses for this subject. Contact the author if you wish to do this course or other courses of the ISSC.
e-mail addresses of the author: `steebwilli@gmail.com`
Home page of the author: `http://issc.uj.ac.za`

Contents

Notation

$:=$	is defined as
$\in$	belongs to (a set)
$\notin$	does not belong to (a set)
$\cap$	intersection of sets
$\cup$	union of sets
$\emptyset$	empty set
$T \subset S$	subset T of set S
$S \cap T$	the intersection of the sets S and T
$S \cup T$	the union of the sets S and T
$\mathbb{N}$	set of natural numbers
$\mathbb{N}_0$	set of natural numbers including 0
$\mathbb{Z}$	set of integers
$\mathbb{Q}$	set of rational numbers
$\mathbb{R}$	set of real numbers
$\mathbb{R}^+$	set of nonnegative real numbers
$\mathbb{C}$	set of complex numbers
$\mathbb{R}^n$	n-dimensional Euclidean space
	space of column vectors with n real components
$\mathbb{C}^n$	n-dimensional complex linear space
	space of column vectors with n complex components
$\mathcal{H}$	Hilbert space
i	$\sqrt{-1}$
$\lfloor x \rfloor$	floor, the greatest integer $\leq x$
$\{x\}$	$x - \lfloor x \rfloor$ the fractional part of x
$\Re(z)$	real part of the complex number z

$\Im(z)$	imaginary part of the complex number z
$\lvert z\rvert$	modulus of complex number z
	$\lvert x+iy\rvert = (x^2+y^2)^{1/2}$, $x, y \in \mathbb{R}$
$f(S)$	image of set S under mapping f
$f^{(n)}$	n-th iterate of the map f
$f \circ g$	composition of two mappings $(f \circ g)(x) = f(g(x))$
x^*	fixed point of map f, i.e. $x^* = f(x^*)$
δ	Dirac delta function
$\mathbf{x}$	column vector in $\mathbb{C}^n$
$\mathbf{x}^T$	transpose of $\mathbf{x}$ (row vector)
$\mathbf{0}$	zero (column) vector
$\lVert . \rVert$	norm
$\mathbf{x} \cdot \mathbf{y} \equiv \mathbf{x}^* \mathbf{y}$	scalar product (inner product) in $\mathbb{C}^n$
$\mathbf{x} \times \mathbf{y}$	vector product in $\mathbb{R}^3$
$\wedge$	exterior product (wedge product, Grassmann product)
$\otimes$	Kronecker product
A, B, C	$m \times n$ matrices
AB	matrix product of $m \times n$ matrix A
	and $n \times p$ matrix B
$\det(A)$	determinant of a square matrix A
tr(A)	trace of a square matrix A
A^*	conjugate transpose of matrix A
A^{-1}	inverse of square matrix A (if it exists)
I_n	$n \times n$ unit matrix
0_n	$n \times n$ zero matrix
δ_{jk}	Kronecker delta with $\delta_{jk} = 1$ for $j = k$
	and $\delta_{jk} = 0$ for $j \neq k$
λ	eigenvalue
r	real bifurcation parameter

Chapter 1

One-Dimensional Maps

1.1 Notations and Definitions

We consider exercises for nonlinear one-dimensional maps. In particular we consider one-dimensional maps with chaotic behaviour. We first summarize the relevant definitions such as fixed points, stability, periodic orbit, Liapunov exponent, invariant density, topologically conjugacy, etc.. Ergodic maps are also considered.

We use the notation $f : D \to C$ to indicate a function f with domain D and codomain C. The notation $f : D \to D$ indicates that the domain and codomain of the function are the same set.

We also use the following two definitions: A mapping $g : A \mapsto B$ is called *surjective* if $g(A) = B$. A mapping g is called *injective* (one-to-one) when $\forall a, a' \in A, g(a) = g(a') \Rightarrow a = a'$. If the mapping f is surjective and injective, the mapping f is called bijective.

Definition. If $B \subset C$, then $f^{(-1)}(B)$ is called the *inverse image* or *preimage* of B and consists of all elements of D whose image is contained in B. That is

$$f^{(-1)}(B) := \{ x \in D : f(x) \in B \}.$$

Note that the notation $f^{(-1)}$ does not necessarily imply that f is an invertible function.

Definition. Consider a map $f : S \to S$. A point $x^* \in S$ is called a *fixed point* of f if

$$f(x^*) = x^*$$

Definition. Let $f : A \to A$ and $g : B \to B$ be two maps. The maps f and g are said to be *topologically conjugate* if there exists a homeomorphism $h : A \to B$ such that, $h \circ f = g \circ h$.

Definition. Consider a map $f : S \to S$. A point $x \in S$ is an eventually fixed point of the function, if there exists $N \in \mathbb{N}$ such that

$$f^{(n+1)}(x) = f^{(n)}(x)$$

whenever $n \geq N$. The point x is eventually periodic with period k, if there exists N such that $f^{(n+k)}(x) = f^{(n)}(x)$ whenever $n \geq N$.

Definition. Let f be a function and p be a periodic point of f with prime period k. Then the point x is *forward asymptotic* to p if the sequence

$$x, \quad f^{(k)}(x), \quad f^{(2k)}(x), \quad f^{(3k)}(x), \ldots$$

converges to p. In other words,

$$\lim_{n \to \infty} f^{(nk)}(x) = p.$$

Definition. The stable set of p, denoted by $W^s(p)$, consists of all points which are forward asymptotic to p. If the sequence

$$|x|, \quad |f(x)|, \quad |f^{(2)}(x)|, \quad |f^{(3)}(x)|, \ldots$$

grows without bound, then x is forward asymptotic to ∞. The stable set of ∞, denoted by $W^s(\infty)$, consists of all points which are forward asymptotic to ∞.

Definition. Let p be a periodic point of the differentiable function f with prime period k. Then p is a *hyperbolic periodic point* if

$$\left| \frac{df^{(k)}}{dx}(x = p) \right| \neq 1.$$

If

$$\left| \frac{df^{(k)}}{dx}(x = p) \right| = 1$$

then p is a nonhyperbolic periodic point.

Definition. Let f be a map of an interval into itself. Consider the one-dimensional difference equation

$$x_{t+1} = f(x_t), \qquad t = 0, 1, 2, \ldots$$

with chaotic behaviour. Assume that in its chaotic regime the map f has a unique invariant measure which is absolutely continuous with respect to the Lebesgue measure. By virtue of ergodicity, the invariant density, denoted by ρ, is determined as a unique solution to the equation

$$\rho(x) = \int_I \delta(x - f(y))\rho(x)\, dy.$$

This equation is called the *Frobenius-Perron integral equation.*

Definition. Consider one-dimensional maps $f : I \to I$. We assume that f is differentiable. One defines the average rate of growth as

$$\lambda(x_0, \delta x_0) = \lim_{n\to\infty} \frac{1}{n} \ln |D_{x(0)} f^{(n)} \delta x(0)|$$

where δx satisfies the variational equation. By a theorem of Oseledec, this limit exists for almost all $x(0)$ with respect to the invariant measure. The average expansion value depends on the direction of the initial perturbation $\delta x(0)$, as well on $x(0)$. If the invariant measure is ergodic, the largest λ with respect to changes of $\delta x(0)$ is independent of $x(0)$, μ-almost everywhere. The number λ_1 is called the largest Liapunov exponent of the map f with respect to the measure μ.

Definition. The *topological entropy*, $H(f)$ gives a measure of the number of distinct trajectories generated by a map f. The topological entropy is a property of f alone and is not associated with any metric properties of the dynamics. It provides a measure of the number of trajectories, or orbits,

$$\{x, f(x), f^{(2)}(x) \ldots\}$$

the map f has. This appears to be infinite, like the number of choices for x. However, orbits

$$\{x, f(x), f^{(2)}(x) \ldots\}$$

and

$$\{y, f(y), f^{(2)}(y) \ldots\}$$

are only considered distinct if

$$|f^{(k)}(x) - f^{(k)}(y)| > \epsilon \text{ for some } k > 0.$$

If one observes up to the nth iterate there will only exist an enumerable number of orbits. If $m(\epsilon, n)$ is the maximum number of different orbits (that is trajectories separated by greater than ϵ) of length n, the topological entropy is defined as a measure of the exponential growth of M with n in the limit of arbitrarily fine discrimination between trajectories, namely

$$H(f) \equiv \lim_{\epsilon \to 0} \lim_{n \to \infty} \frac{1}{n} \ln(M(n, \epsilon)).$$

This indicates $M \sim \exp(Hn)$ in the limit. The topological entropy gives the rate of growth of orbits with finite length as their allowed length goes to infinity ($n \to \infty$) and resolution fidelity becomes arbitrarily fine ($\epsilon \to 0$).

Let f be a one-dimensional map. The *topological entropy* is also determined by the number of fixed points of $f^{(n)}$, namely

$$H(f) := \lim_{n \to \infty} \frac{1}{n} \ln(\text{number of fixed points under the map} f^{(n)}).$$

1.2 One-Dimensional Maps

1.2.1 Solved Problems

Problem 1. Consider the analytic function $f : \mathbb{R} \to \mathbb{R}$

$$f(x) = 4x(1-x)$$

or written as difference equation

$$x_{t+1} = 4x_t(1 - x_t)$$

with $t = 0, 1, \ldots$ and $x_0 \in \mathbb{R}$.
(i) The *fixed points* of the function f are the solutions of the equation $f(x^*) = x^*$. Find the fixed points.
(ii) Find the *variational equation* (*linearized equation*)

$$y_{t+1} = \frac{df}{dx}(x = x_t) y_t$$

and study the stability of the fixed points found in (i).
(iii) The *critical points* of f are the solutions of the equation $df(x)/dx = 0$. Find the critical points of f. If there are critical points determine whether they relate to minima or maxima of f.
(iv) The *roots* of the function f are the solutions of $f(x) = 0$. Find the roots of f.
(v) Find the fixed points of the analytic function $g(x) = f(f(x))$.

(vi) Find the critical points of the analytic function $g(x) = f(f(x))$. If there are critical points of g determine whether they relate to minima or maxima.
(vii) Find the roots of the analytic function $g(x) = f(f(x))$.

If $x \in [0,1]$, then $f(x) \in [0,1]$. So the function f could be restricted to $f : [0,1] \to [0,1]$. For the difference equation $x_{t+1} = 4x_t(1-x_t)$ we have $x_0 \in [0,1]$.

Solution 1. (i) Solving the quadratic equation $4x^*(1-x^*) = x^*$ provides the two fixed points

$$x^* = 0, \qquad x^* = \frac{3}{4}.$$

(ii) Since $df/dx = 4 - 8x$ we obtain the variational equation

$$y_{t+1} = (4 - 8x_t)y_t.$$

Inserting the fixed point $x^* = 0$ we obtain $y_{t+1} = 4y_t$ with the solution $y_t = 4^t y_0$. Thus the fixed point is unstable. Inserting the fixed point $x^* = 3/4$ we obtain $y_{t+1} = -2y_t$ with the solution $y_t = (-2)^t y_0$. Thus the fixed point $x^* = 3/4$ is unstable.
(iii) The derivative of f is

$$\frac{df(x)}{dx} = 4 - 8x.$$

From $4 - 8x = 0$ we find the critical point $x = 1/2$. Since $d^2f/dx^2 = -8$ the critical point $x = 1/2$ refers to a maximum.
(iv) Solving the quadratic equation $4x(1-x) = 0$ provides the roots $x = 0$ and $x = 1$.
(v) We have

$$g(x) = f(f(x)) = 16x(1-x)(1-4x(1-x)).$$

The fixed point equation is $g(x^*) = x^*$. The fixed points of f are also fixed points of g. Thus we can reduce the quartic equation $g(x^*) = x^*$ to a quadratic equation and find the four fixed points

$$x^* = 0, \quad x^* = \frac{3}{4}, \quad x^* = \frac{5+\sqrt{5}}{8}, \quad x^* = \frac{5-\sqrt{5}}{8}.$$

All of them lie in the unit interval $[0,1]$.
(vi) The derivative of $g(x) = f(f(x))$ is given by

$$\frac{dg(x)}{dx} = -256x^3 + 384x^2 - 160x + 16.$$

The solutions of the cubic equation $dg/dx = 0$ are

$$\frac{1}{2}, \quad \frac{2-\sqrt{2}}{4}, \quad \frac{2+\sqrt{2}}{4}.$$

All of them lie in the unit interval $[0,1]$. The second derivative of g is

$$\frac{d^2g(x)}{dx^2} = -768x^2 + 768x - 160.$$

Inserting the critical point $(2-\sqrt{2})/4$ into d^2g/dx^2 yields -64, i.e. we have a maximum. Inserting the critical point $(2+\sqrt{2})/4$ into d^2g/dx^2 yields -64, i.e. we have a maximum. Inserting the critical point $1/2$ into d^2g/dx^2 yields $+32$, i.e. we have a minimum.
(vii) Solving the quartic equation $g(x) \equiv f(f(x)) = 0$ provides the roots

$$0, \quad 1, \quad \frac{1}{2} \text{ twice}$$

The roots 0 and 1 we already found for the function f.

Problem 2. Consider the logistic map $f : [0,1] \to [0,1]$

$$f(x) = 4x(1-x).$$

The fixed points are $x^* = 0$ and $x^* = 3/4$. Both fixed points are unstable.
(i) Let $\widetilde{x} = 3/4$. Find the *preimage.*
(ii) Let $\widetilde{x} = 0$. Find the preimage.
(iii) Let $\widetilde{x} = 1/2$. Find the preimage.
(iv) Let n be an integer with $n \geq 3$. Show that

$$\left| \frac{1}{n} - \frac{1}{n+1} \right| < |f(1/n) - f(1/(n+1))|.$$

Solution 2. (i) We have to solve $f(x) = \widetilde{x}$, i.e. $4x(1-x) = \widetilde{x} = \frac{3}{4}$. This quadratic equation provides two solutions $x_1 = 3/4$ and $x_2 = 1/4$. Note that $x_1 = 3/4$ is also a fixed point of the map f.
(ii) We have to solve $f(x) = \widetilde{x}$, i.e. $4x(1-x) = \widetilde{x} = 0$. This quadratic equation provides two solutions $x_1 = 0$ and $x_2 = 1$. Note that $x_1 = 0$ is a fixed point of f.
(iii) We have to solve $f(x) = \widetilde{x}$, i.e. $4x(1-x) = \widetilde{x} = 1/2$. This quadratic equation provides the two solutions

$$x_1 = \frac{1}{2} + \frac{1}{2\sqrt{2}}, \quad x_2 = \frac{1}{2} - \frac{1}{2\sqrt{2}}.$$

(iv) We have

$$\frac{1}{n} - \frac{1}{n+1} \equiv \frac{1}{n(n+1)} \equiv \frac{n(n+1)}{n^2(n+1)^2}$$

and

$$f(1/n) - f(1/(n+1)) = 4\frac{n^2-n-1}{n^2(n+1)^2}.$$

Since $n(n+1) < 4(n^2-n-1)$ for all $n \geq 3$ the result follows.

Problem 3. (i) Let $f : \mathbb{R} \to \mathbb{R}$, $g : \mathbb{R} \to \mathbb{R}$ be analytic functions with the same fixed point x^*, i.e. $f(x^*) = x^*$, $g(x^*) = x^*$. Show that $f \circ g$ and $g \circ f$ admit this fixed point, where $\circ$ denotes function composition.
(ii) Let $f_1 : \mathbb{R} \to \mathbb{R}$, $f_2 : \mathbb{R} \to \mathbb{R}$ be continuous functions. Assume that $x^* = 0$ is a fixed point of both f_1 and f_2. Let $g_1(x) = f_1(f_2(x))$ and $g_2(x) = f_2(f_1(x))$. Show that the functions $h_1(x) = g_1(x) - g_2(x)$ and $h_2(x) = g_1(x) + g_2(x)$ also admit this fixed point.

Solution 3. (i) We have

$$f(g(x^*)) = f(x^*) = x^*, \qquad g(f(x^*)) = g(x^*) = x^*.$$

(ii) We have

$$h_1(0) = f_1(f_2(0)) - f_2(f_1(0)) = f_1(0) - f_2(0) = 0 - 0 = 0$$

and

$$h_2(0) = f_1(f_2(0)) + f_2(f_1(0)) = f_1(0) + f_2(0) = 0 + 0 = 0.$$

Problem 4. Let $f : \mathbb{R} \to \mathbb{R}$ with $f(x) = 4x(1-x)$. Then $df/dx = 4 - 8x$. Study the two-dimensional map

$$x_{t+1} = 4x_t(1-x_t), \quad y_{t+1} = |(4-8x_t)|y_t$$

where $x_0, y_0 \in (0,1)$. Let $x_0 = 1/3$ and $y_0 = 1/2$. Find x_1, x_2 and y_1, y_2.

Solution 4. We find

$$x_1 = \frac{8}{9}, \qquad x_2 = \frac{32}{81}$$

and

$$y_1 = \frac{2}{3}, \qquad y_2 = \frac{56}{27}.$$

The quantity y_t from the variational equation is utilized to calculate the *Liapunov exponent*.

Problem 5. The logistic map is given by

$$x_{t+1} = 4x_t(1 - x_t), \qquad t = 0, 1, \ldots \tag{1}$$

where $x_0 \in [0, 1]$. The fixed points are given by $x^* = 0$, $x^* = 3/4$.
(i) Show that $x_t \in [0, 1]$ for $t = 1, 2, \ldots$.
(ii) Give the variational equation.
(iii) Show that the fixed points are unstable. Hint. Show that

$$\left| \frac{df(x)}{dx} \right|_{x=x^*} > 1. \tag{2}$$

(iv) Find the periodic orbits.
(v) Show that the exact solution of (1) is given by

$$x_t = \frac{1}{2} - \frac{1}{2} \cos(2^t \arccos(1 - 2x_0))$$

where $x_0 \in [0, 1]$ is the initial value.
(vi) Show that for almost all initial values the Liapunov exponent λ for the logistic map (1) is given by $\lambda = \ln(2)$.
(vii) The *time average* is defined by

$$\langle x_t \rangle := \lim_{T \to \infty} \frac{1}{T} \sum_{t=0}^{T} x_t.$$

For almost all initial values we find $\langle x_t \rangle = 1/2$. The *autocorrelation function* is defined by

$$C_{xx}(\tau) := \lim_{T \to \infty} \frac{1}{T} \sum_{t=0}^{T} (x_t - \langle x_t \rangle)(x_{t+\tau} - \langle x_t \rangle)$$

where $\tau = 0, 1, 2, \ldots$. Find the autocorrelation function.

Solution 5. (i) Let $x \in [0, 1]$. Then $x(1 - x) \leq 1/4$. Consequently, $4x(1 - x) \leq 1$. If $x \neq \frac{1}{2}$, then $x(1 - x) < \frac{1}{4}$. In other words, the function $g(x) = x(1 - x)$ has one maximum at $x = \frac{1}{2}$ with $g(\frac{1}{2}) = \frac{1}{4}$.
(ii) Let $x_{t+1} = f(x_t)$ be a one-dimensional difference equation. Assume that f is differentiable. Then

$$y_{t+1} = \frac{df}{dx}(x_t) y_t$$

is called the (difference) variational equation or linearized equation. Since

$$\frac{df(x)}{dx} = 4 - 8x$$

we obtain the variational equation $y_{t+1} = (4-8x_t)y_t$, where $t = 0, 1, 2, \ldots$.
(iii) Inserting the fixed point $x^* = 0$ into the variational equation yields $y_{t+1} = 4y_t$. The solution is given by $y_t = 4^t y_0$. Therefore the fixed point $x^* = 0$ is unstable. We have $|df(0)/dx| > 1$ for the fixed point $x^* = 0$. Inserting the fixed point $x^* = 3/4$ into the variational equation yields $y_{t+1} = -2y_t$. Therefore this fixed point is also unstable. We have $|df(3/4)/dx| > 1$ for the fixed point $x^* = 3/4$.
(iv) The periodic orbits are given by the initial values

$$x_0 = \frac{1}{2} - \frac{1}{2}\cos\left(\frac{r\pi}{2^s}\right)$$

where r and s are positive integers. We have

$$\arccos(1-2x_0) = \arccos\left(\cos\left(\frac{r\pi}{2^s}\right)\right) = \frac{r\pi}{2^s}.$$

It follows that

$$x_t = \frac{1}{2} - \frac{1}{2}\cos\left(\frac{2^t r\pi}{2^s}\right).$$

(v) Let $\alpha := \cos^{-1}(1-2x_0)$. Then

$$x_t = \frac{1}{2} - \frac{1}{2}\cos(2^t\alpha).$$

It follows that

$$x_{t+1} = \frac{1}{2} - \frac{1}{2}\cos(2^{t+1}\alpha) = \frac{1}{2} - \frac{1}{2}\left(2\cos^2(2^t\alpha) - 1\right) = 1 - \cos^2(2^t\alpha).$$

Now

$$4x_t(1-x_t) = 4x_t - 4x_t^2 = 2 - 2\cos(2^t\alpha) - 1 - \cos^2(2^t\alpha) + 2\cos(2^t\alpha).$$

Therefore

$$4x_t(1-x_t) = 1 - \cos^2(2^t\alpha).$$

This proves that x_t is the general solution of equation (1), where x_0 is the initial value.
(vi) The *Liapunov exponent* λ is given by

$$\lambda(x_0) := \lim_{T\to\infty}\frac{1}{T}\sum_{t=0}^{T-1}\ln\left|\frac{df(x)}{dx}\right|_{x=x_t}$$

where x_0 is the initial value, i.e. the Liapunov exponent depends on the initial value. The Liapunov exponent measures the exponential rate at which the derivative grows. Since $df/dx = 4-8x$ we find that the Liapunov exponent is given by

$$\lambda(x_0) = \ln(2)$$

for almost all initial values, i.e. the periodic orbits given above are excluded.
(vii) Using the exact solution

$$x_t = \frac{1}{2} - \frac{1}{2}\cos(2^t \cos^{-1}(1 - 2x_0))$$

we find that the autocorrelation function is given by

$$C_{xx}(\tau) = \begin{cases} \frac{1}{8} & \text{for} \quad \tau = 0 \\ 0 & \text{otherwise} \end{cases}$$

Problem 6. Consider the logistic map

$$x_{t+1} = 4x_t(1 - x_t), \qquad t = 0, 1, 2, \dots \ .$$

Apply the transformation $x_t = \sin^2(\pi\theta_t)$, i.e. find the map for θ_t, where $\theta_t \in [0, 1)$.

Solution 6. Using the identity $\sin(2\alpha) \equiv 2\sin(\alpha)\cos(\alpha)$ we obtain

$$\begin{aligned} \sin^2(\pi\theta_{t+1}) &= 4\sin^2(\pi\theta_t)(1 - \sin^2(\pi\theta_t)) = 4\sin^2(\pi\theta_t)\cos^2(\pi\theta_t) \\ &= (2\sin(\pi\theta_t)\cos(\pi\theta_t))^2 \\ &= \sin^2(2\pi\theta_t). \end{aligned}$$

Thus $\theta_{t+1} = 2\theta_t \mod 1$.

Problem 7. Let $f : \mathbb{R} \to \mathbb{R}$ be a continuously differentiable map. Let $f^{(n)}$ be the n-th iterate of f.
(i) Calculate the derivative of $f^{(n)}$ at x_0.
(ii) Let $f(x) = 2x(1 - x)$. Find $df^{(2)}/dx$.

Solution 7. (i) Applying the *chain rule* we find

$$\frac{df^{(n)}(x)}{dx} = f'(f^{(n-1)}(x))f'(f^{(n-2)}(x)) \cdots f'(f(x))f'(x).$$

With x_0, $x_1 = f(x_0)$, $x_2 = f(f(x_0)) = f(x_1)$, ... we can write

$$\frac{df^{(n)}}{dx}(x_0) = f'(x_{n-1})f'(x_{n-2}) \cdots f'(x_1)f'(x_0).$$

(ii) Since $f'(x) = 2 - 4x$ and $f'(f(x)) = 2 - 4(2x - x^2)$ we have

$$\frac{df^{(2)}(x)}{dx} = f'(f(x))f'(x) = (2 - 4(2x - x^2))(2 - 4x) = 4 - 24x + 48x^2 - 32x^3.$$

Problem 8. Consider the analytic function $f : \mathbb{R} \to \mathbb{R}$, $f(x) = 2x(1-x)$.
(i) Find the fixed points. Are the fixed points stable?
(ii) Calculate

$$\lim_{n\to\infty} f^{(n)}(1/3).$$

Discuss this result.
(iii) Let n be a positive integer with $n \geq 2$. Find the distances

$$|1/n - 1/(n+1)| \quad \text{and} \quad |f(1/n) - f(1/(n+1))|.$$

Compare the distances.

Solution 8. (i) The fixed points are the solutions of the equation $f(x^*) = x^*$. Thus $2x^*(1-x^*) = x^*$ with the solutions $x_1^* = 0$ and $x_2^* = 1/2$. The fixed point $x_1^* = 0$ is unstable since

$$\frac{df(x)}{dx} = 2 - 4x$$

and thus $df(x = x_1^*)/dx = 2$. The fixed point $x_2^* = 1/2$ is stable since

$$\frac{df(x)}{dx} = 2 - 4x$$

and thus $df(x = x_2^*)/dx = 0$.
(ii) We have

$$f(1/3) = 4/9, \quad f(4/9) = 40/81, \quad f(40/81) = 3280/6561.$$

This generates the sequence

$$s_n = \frac{1}{2}\frac{9^{2^{n-1}} - 1}{9^{2^{n-1}}} \quad \text{with} \quad \lim_{n\to\infty} s_n = \frac{1}{2}.$$

This is one of the fixed points of the map.
(iii) We have

$$\left|\frac{1}{n} - \frac{1}{n+1}\right| = \frac{1}{|n(n+1)|}.$$

Since

$$f\left(\frac{1}{n}\right) = \frac{2}{n}\left(1 - \frac{1}{n}\right), \quad f\left(\frac{1}{n+1}\right) = \frac{2}{n+1}\left(1 - \frac{1}{n+1}\right).$$

It follows that

$$\left|\frac{1}{n} - \frac{1}{n+1}\right| > |f(1/n) - f(1/(n+1))|.$$

Problem 9. Consider the analytic function $f : \mathbb{R} \to \mathbb{R}$

$$f(x) = -\frac{3}{2}x^2 + \frac{5}{2}x + 1.$$

(i) Find the fixed points.
(ii) Show that $\{0, 1, 2\}$ form an orbit of period three.

Solution 9. (i) Solving the quadratic equation $-3x^{*2}/2 + 5x^*/2 + 1 = x^*$ we obtain the two fixed points

$$x_1^* = \frac{1}{2} + \frac{\sqrt{11}}{2\sqrt{3}}, \quad x_2^* = \frac{1}{2} - \frac{\sqrt{11}}{2\sqrt{3}}.$$

Are the fixed points stable?
(ii) Let $x = 0$. Then we have

$$f(0) = 1, \quad f(f(0)) = 2, \quad f(f(f(0))) = 0.$$

Problem 10. Consider an integer sequence x_t $(t = 1, 2, \ldots)$ given by

$$x_{2t} = 2x_t - 1 \quad \text{for } t \geq 1$$
$$x_{2t+1} = 2x_t + 1 \quad \text{for } t \geq 1$$

with the initial condition $x_1 = 1$. Find x_2, x_3, $\ldots$, x_{16} and the solution.

Solution 10. We have

$$x_2 = 1, \; x_3 = 3$$

$$x_4 = 1, \; x_5 = 3, \; x_6 = 5, \; x_7 = 7$$

$$x_8 = 1, \; x_9 = 3, \; x_{10} = 5, \; x_{11} = 7, \; x_{12} = 9, \; x_{13} = 11, \; x_{14} = 13, \; x_{15} = 15$$

and $x_{16} = 1$. The solution is

$$x_{2^m+k} = 2k + 1 \quad \text{for } m \geq 0 \text{ and } 0 \leq k < 2^m.$$

Problem 11. Let $f : \mathbb{Z} \to \mathbb{Z}$ with $f(n) = |n - 1|$.
(i) Show that 0 and 1 form a periodic cycle.
(ii) Show that $n = 2$ is eventually periodic.
(iii) Show that every integer is eventually periodic.
(iv) Does the map f admit a fixed point?

Solution 11. (i) We have $f(0) = 1$, $f(1) = 0$.

(ii) Let $n = 2$. Then

$$f(2) = 1, \quad f(f(2)) = f(1) = 0, \quad f(f(f(2))) = f(f(1)) = f(0) = 1.$$

(iii) Since $f(n) = n - 1$ for $n \in \mathbb{Z}$, $n > 1$ and $f(n) = |n| + 1$ for $n \in \mathbb{Z}$, $n < 1$ we find that

$$f^{(n)}(n) = 0, \quad n > 1$$

$$f^{(|n|+2)}(n) = 0, \quad n < 1.$$

Thus f is eventually periodic for $n \in \mathbb{Z}$.
(iv) The equation $|n - 1| = n$ admits no solution.

Problem 12. Let $g : \mathbb{Z} \to \mathbb{Z}$ with $g(n) = |n^2 - 1|$.
(i) Find $g(0)$, $g(g(0))$. Discuss.
(ii) Find the fixed points of g. Find the fixed points of $g \circ g$.

Solution 12. (i) We have $g(0) = 1$ and $g(g(0)) = 0$. Thus we have a periodic orbit.
(ii) The equation $|n^2 - 1| = n$ has no solution so there are no fixed points of g. Owing to the result of (i) $g(g(n))$ has the two fixed points 0 and 1.

Problem 13. (i) Consider the *negation map* $f : \mathbb{R} \to \mathbb{R}$, i.e. $f(x) = -x$. Find all fixed points. Find all its periodic points.
(ii) Let $g : \mathbb{R} \to \mathbb{R}$ with $g(x) = -x^3$. Show that g has a fixed point at 0 and a periodic cycle consisting of 1 and -1. Find the stable set.
(iii) Consider the cubic map $h : \mathbb{R} \to \mathbb{R}$ with $h(x) = x^3$. Find all its periodic points and the stable set of each. Calculate the iterate $h^{(n)}(x)$. Assume that $x \in (-1, 1)$. Find $\lim_{n\to\infty} h^{(n)}(x)$.

Solution 13. (i) The equation $-x^* = x^*$ admits the only solution $x^* = 0$ as a fixed point. Iteration of the map yields the sequence $x, -x, x, -x, \ldots$. Hence the map has a fixed point at 0 and all other points on the real line are periodic points of prime period 2.
(ii) Obviously, $x^* = 0$ is a fixed point, since $g(x^*) = x^*$. This is the only fixed point of the map. The fixed point is stable. The stable set of 0 consists of all points in the interval $(-1, 1)$. The stable set of ∞ consists of all points in the interval $(-\infty, -1)$ and $(1, \infty)$. The periodic points are 1 and -1. The stable set of $\{-1\}$ contains only the point -1. The map g has no other periodic points.
(iii) Periodic points of period n are found from the equation

$$p = h^{(n)}(p). \tag{1}$$

By iteration one finds

$$h^{(n)}(p) = h^{(n-1)}(p^3) = h^{(n-2)}(p^9) = \cdots = p^{3^n}.$$

Substituting $p = h^{(n)}(p)$ into this equation yields

$$p(p^{3^n-1} - 1) = 0.$$

Since $3^n - 1$ is even, the only real roots of this equation are $0, \pm 1$. These points are all fixed points. It follows from (1) that $|h^{(n)}(x)| \to 0$ if $n \to \infty$ for $|x| < 1$ and $|h^{(n)}(x)| \to \infty$ if $n \to \infty$ for $|x| > 1$. Therefore

$$W^s(0) = (-1, 1), \quad W^s(1) = \{1\}, \quad W^s(-1) = \{-1\}$$

$$W^s(\infty) = (-\infty, -1) \cup (1, \infty).$$

We obtain

$$h^{(n)}(x) = x^{3^n}.$$

Thus for $x \in (-1, 1)$ we obtain

$$\lim_{n \to \infty} h^{(n)}(x) = 0$$

i.e. $h^{(n)}$ tends to the fixed point 0.

Problem 14. Consider the continuous function $f : \mathbb{R} \to \mathbb{R}$

$$f(x) = x|x|.$$

(i) Is the function differentiable? If so find the derivative.
(ii) Find the fixed points of the function and study their stability.

Solution 14. (i) The first derivative exists with $df/dx = 2|x|$, whereas the second order derivative of f does not exist at $x = 0$.
(ii) The fixed points are the solutions of

$$x^*|x^*| = x^*.$$

Thus we find the three fixed points $x^* = -1, 0, 1$. The fixed point $x^* = -1$ is unstable since $df(-1)/dx = 2$. The fixed point $x^* = 0$ is stable since $df(0)/dx = 0$. The fixed point $x^* = 1$ is unstable since $df(1)/dx = 2$.

Problem 15. (i) Show that the analytic function $f : \mathbb{R} \to \mathbb{R}$,

$$f(x) = \sin(x)$$

admits the fixed point $x^* = 0$. Study the stability of this fixed point.

(ii) Show that the analytic function $f : \mathbb{R} \to \mathbb{R}$,

$$f(x) = \sinh(x)$$

admits the fixed point $x^* = 0$. Study the stability of this fixed point.
(iii) Show that the analytic function $f : \mathbb{R} \to \mathbb{R}$,

$$f(x) = \tanh(x)$$

admits the fixed point $x^* = 0$. Study the stability of this fixed point.

Hint. Let $f : \mathbb{R} \to \mathbb{R}$ be an analytic function and x^* be a fixed point, i.e. $x^* = f(x^*)$. The fixed point x^* is called non-hyperbolic if

$$|f'(x = x^*)| = 1$$

where $'$ denotes derivative. If $f'(x = x^*) = 1$, then there are three cases: (a) If $f''(x = x^*) \neq 0$, then the fixed point x^* is semi-asymptotically from the left if $f''(x = x^*) > 0$ and from the right if $f''(x = x^*) < 0$. (b) If $f''(x = x^*) = 0$ and $f'''(x = x^*) < 0$, then the fixed point x^* is asymptotically stable. (c) If $f''(x = x^*) = 0$ and $f'''(x = x^*) > 0$, then x^* is unstable.
If $f'(x = x^*) = -1$, then two cases have to be studied: (i) If $Sf(x = x^*) < 0$, then the fixed point is asymptotically stable. (ii) If $Sf(x = x^*) > 0$, then the fixed point x^* is unstable. Here $Sf(x)$ denotes the *Schwarzian derivative* defined by

$$Sf(x) := \frac{f'''(x)}{f''(x)} - \frac{3}{2}\left(\frac{f''(x)}{f'(x)}\right)^2.$$

Solution 15. (i) Since $\sin(0) = 0$ we find that $x^* = 0$ is a fixed point. Now the derivative is

$$\frac{df(x)}{dx} = \cos(x)$$

with $\cos(0) = 1$. The *Taylor expansion* for $\sin(x)$ around 0 is

$$\sin(x) = x - \frac{x^3}{3!} + \frac{x^5}{5!} - \frac{x^7}{7!} + \cdots$$

Note that $\sin(x) < x$ for $x \in (-\pi/2, \pi/2)$. Thus the fixed point $x^* = 0$ is stable.
(ii) Since $\sinh(0) = 0$ we find that $x^* = 0$ is a fixed point. The derivative is

$$\frac{df(x)}{dx} = \cosh(x)$$

with $\cosh(0) = 1$. The Taylor expansion of $\sinh(x)$ around 0 is

$$\sinh(x) = x + \frac{x^3}{3!} + \frac{x^5}{5!} + \cdots$$

Thus the fixed point is unstable.
(iii) Since $\tanh(0) = 0$ we find that $x^* = 0$ is a fixed point. The derivative is

$$\frac{df(x)}{dx} = \frac{1}{\cosh(x)}$$

with $1/\cosh(0) = 1$. The expansion of $\tanh(x)$ around 0 is

$$\tanh(x) = x - \frac{x^3}{3} + \frac{2x^5}{16} - \frac{17x^7}{315} + \cdots \qquad |x| < \frac{\pi}{2}.$$

Thus the fixed point $x^* = 0$ is stable. Note that

$$\frac{d^2 f(x)}{dx^2} = -\sinh(x)/\cosh^2(x)$$

with $d^2 f(0)/dx^2 = 0$ and $d^3 f(x)/dx^3 = (-\cosh^2(x) + 2\sinh^2(x))/\cosh^3(x)$ with $d^3 f(0)/dx^3 < 0$.

Problem 16. Let $f : \mathbb{R} \to \mathbb{R}$ be an analytic function. A fixed point x^* is called hyperbolic if $|f'(x = x^*)| \neq 1$ and non-hyperbolic if $|f'(x = x^*)| = 1$. Let S be the *Schwarzian derivative*

$$Sf(x) := \frac{f'''(x)}{f'(x)} - \frac{3}{2}\left(\frac{f''(x)}{f'(x)}\right)^2.$$

If $f'(x = x^*) = -1$ we have two cases: (a) If $Sf(x = x^*) < 0$, then x^* is asymptotically stable, (b) If $Sf(x = x^*) > 0$, then x^* is unstable. Apply it to the analytic function $f(x) = -\sinh(x)$.

Solution 16. The only fixed point is $x^* = 0$. Since $f'(x) = -\cosh(x)$, $f''(x) = -\sinh(x)$, $f'''(x) = -\cosh(x)$ we obtain

$$Sf(x) = 1 - \frac{3}{2}\frac{\sinh(x)}{\cosh(x)}.$$

Hence $Sf(0) > 0$ and the fixed point $x^* = 0$ is unstable.

Problem 17. Give examples of maps $f : [0,1] \to [0,1]$ where f is 1 to 1 and monotone on the interval $[0, 0.5]$ and satisfy the conditions

$$f(0) = 0, \quad f(0.5) = 1, \quad f(1) = 0, \quad f(x) = f(1-x).$$

Solution 17. Examples are the *logistic map*

$$f_\ell(x) = 4x(1-x)$$

the symmetric *tent map*

$$f_t(x) = 1 - 2|x - 0.5|$$

the *sine map*

$$f_s(x) = \sin(\pi x)$$

the *entropy map*

$$f_e(x) = -x\log_2(x) - (1-x)\log_2(1-x)$$

the *Bell map*

$$f_b(x) = \frac{e^{-(x-0.5)^2} - e^{-0.25}}{1 - e^{-0.25}}.$$

Obviously all admit the fixed point $x^* = 0$ and one other fixed point in the domain $(0.5, 1)$.

Problem 18. Can one find polynomials $p : \mathbb{R} \to \mathbb{R}$ such that one critical point of p and one fixed point of p coincide? Start of with

$$p(x) = c_n x^n + c_{n-1}x^{n-1} + \cdots + c_1 x + c_0$$

where $n \geq 2$.

Solution 18. We have

$$\frac{dp}{dx} = c_n n x^{n-1} + c_{n-1}(n-1)x^{n-2} + \cdots + 2c_2 x + c_1.$$

The condition for the critical points is $dp/dx = 0$. Multiplying the condition $dp/dx = 0$ by x yields

$$x\frac{dp}{dx} = c_n n x^n + c_{n-1}(n-1)x^{n-1} + \cdots + 2c_2 x^2 + c_1 x = 0.$$

The equation for the fixed points is

$$p(x) = c_n x^n + c_{n-1}x^{n-1} + \cdots + c_2 x^2 + c_1 x + c_0 = x.$$

Subtracting these two equations provides

$$c_n(n-1)x^n + c_{n-1}(n-2)x^{n-1} + \cdots + c_2 x^2 - c_0 = -x.$$

This equation can only be satisfied if $x = 0$ and $c_0 = 0$. Then from $dp/dx = 0$ it follows that $c_1 = 0$. Thus if $c_0 = c_1 = 0$, then $x = 0$ is a fixed point of p and it is also a critical point of p. It follows that

$$\frac{dp}{dx}(x = 0) = c_1.$$

Problem 19. Let $b > a$. Consider a differentiable map $f : [a, b] \to [a, b]$. A fixed point x^* of the map f is called *superstable* if $df(x = x^*)/dx = 0$. Let $r \in [1, 4]$ and $f_r : [0, 1] \to [0, 1]$

$$f_r(x) = rx(1 - x).$$

Does the function f_r for some $r \in [1, 4]$ admit a superstable fixed point?

Solution 19. We have the fixed points $x^* = 0$ and $x^* = (r - 1)/r$. Now the derivative of f_r is given by

$$\frac{df_r(x)}{dx} = r(1 - 2x).$$

Thus for $x^* = 0$ we have $df_r(0)/dx = r$ and thus the fixed point $x^* = 0$ is not superstable. For the fixed point $x^* = (r - 1)/r$ we have

$$\frac{df_r(x = (r - 1)/r)}{dx} = 2 - r.$$

Thus for $r = 2$ we have the superstable fixed point $x^* = 1/2$.

Problem 20. Let $a \neq 0$. Consider the polynomials

$$f(x) = ax^3 + bx^2 + cx + d, \qquad g(x) = x^3 + Ax + B$$

with $A = 9ac - 3b^2$, $B = 27a^2d + 2b^3 - 9abc$.
(i) Find the *Newton maps*

$$N_f(x) := x - \frac{f(x)}{f'(x)}, \qquad N_g(x) := x - \frac{g(x)}{g'(x)}.$$

(ii) Let $h(x) = 3ax + b$. Show that $h \circ N_f = N_g \circ h$.

Solution 20. (i) We find

$$N_f(x) = \frac{2ax^3 + bx^2 - d}{3ax^2 + 2bx + c}, \qquad N_g(x) = \frac{2x^3 - B}{3x^2 + A}.$$

(ii) We have

$$(h \circ N_f)(x) = h(N_f(x)) = \frac{3a(2a^3 + bx^2 - d)}{3ax^2 + 2bx + c} + b$$
$$= \frac{6a^2x^3 + 6abx^2 + 2b^2x - 3ad + bc}{3ax^2 + 2bx + c}$$

and

$$(N_g \circ h)(x) = N_g(h(x)) = \frac{2(3ax + b)^3 - B}{3(3ax + b)^2 + A}$$
$$= \frac{6a^2x^3 + 6abx^2b^2x - 3ad + bc}{3ax^2 + 2bx + c}.$$

Problem 21. (i) Let $r > 0$. Find the first iterate of the *Newton map*

$$f_r(x) = \frac{1}{2}\left(x + \frac{r}{x}\right).$$

(ii) Let $n \in \mathbb{N}$. Consider the map (*Newton's method*) to find the square root of n

$$x_{t+1} = \frac{1}{2}\left(x_t + \frac{n}{x_t}\right), \qquad t = 0, 1, 2, \ldots$$

given the initial value x_0 with $x_0 > 0$. Find the fixed points. Show that

$$\frac{x_{t+1} - \sqrt{n}}{x_{t+1} + \sqrt{n}} = \left(\frac{x_t - \sqrt{n}}{x_t + \sqrt{n}}\right)^2.$$

Find $\lim_{t \to \infty} x_t$.

Solution 21. (i) We have

$$f_r(f_r(x)) = \frac{1}{2}\left(\frac{1}{2}\left(x + \frac{r}{x}\right) + \frac{r}{\frac{1}{2}(x + \frac{r}{x})}\right)$$
$$= \frac{1}{2}\frac{(x^4 + 2rx^2 + r^2)/2 + 2rx^2}{x(x^2 + r)}$$
$$= \frac{1}{2}\frac{x^4/2 + 3rx^2 + r^2/2}{x(x^2 + r)}.$$

(ii) From

$$\frac{1}{2}\left(x^* + \frac{n}{x^*}\right) = x^*$$

we obtain $x^* = n/x^*$ and thus $x^{*2} = n$ with the fixed point $x^* = \sqrt{n}$. We have

$$\frac{\frac{1}{2}\left(x_t + \frac{n}{x_t}\right) - \sqrt{n}}{\frac{1}{2}\left(x_t + \frac{n}{x_t}\right) + \sqrt{n}} = \frac{x_t^2 + n - 2x_t\sqrt{n}}{x_t^2 + n + 2x_t\sqrt{n}} = \frac{(x_t - \sqrt{n})^2}{(x_t + \sqrt{n})^2}.$$

We obtain $\lim_{t\to\infty} x_t = \sqrt{n}$.

Problem 22. Consider the polynomial $f(x) = x^3 - 3x + 3$. Show that for any positive integer T, there is an initial value x_0 such that the sequence $x_0, x_1, x_2, \ldots$ obtained from *Newton's method*

$$x_{t+1} = x_t - \frac{f(x_t)}{f'(x_t)} = \frac{2x_t^3 - 3}{3(x_t^2 - 1)}, \qquad t = 0, 1, 2, \ldots$$

has period T.

Solution 22. For $T = 1$ (i.e. period 1) we need $f(x_0) = 0$. Any cubic polynomial with real coefficients has at least one real root. Choose this real root as x_0. Thus $f(x_0) = 0$. Consider now the case $T \geq 2$. We define the Newton map

$$F(x) := x - \frac{f(x)}{f'(x)} = \frac{2x^3 - 3}{3(x^2 - 1)}$$

with the sequence x_0, $x_1 = F(x_0)$, $x_2 = F(F(x_0))$, ... The derivative of F yields

$$F'(x) = \frac{f''(x)f(x)}{(f'(x))^2} = \frac{6x}{(3(x^2 - 1))^2}(x^3 - 3x + 3).$$

Since f increases on the interval $[1, \infty)$, it is at least 1 there. Thus we find that $F'(x) > 0$ on $(1, \infty)$ and F is increasing on $(1, \infty)$. We have

$$\lim_{x\to 1^+} F(x) = -\infty, \qquad \lim_{x\to\infty} F(x) = +\infty.$$

Hence if F is restricted to $(1, \infty)$ it has a continuous inverse function G whose domain is $(-\infty, \infty)$. Consequently for any real number x, $G(x)$ is the unique number in $(1, \infty)$ for which $F(G(x)) = x$. Let $G^{(k)}$ be the function composition of k copies of G. Define the function H on $(-1, 0]$ by $H(x) := F(x) - G^{(T-1)}(x)$. The functions F, G and H are continuous on $(-1, 0]$. Since $G^{(T-1)}(-1)$ is finite we have

$$\lim_{x\to -1^+} H(x) = +\infty.$$

Now $F(0) = 1$ and all values of G are greater than 1. Consequently $H(0) < 0$. By the intermediate value theorem there exists an $x_0 \in (-1, 0)$ for which

$H(x_0) = 0$. Then the sequence starting from this initial value x_0 has period T. Since $H(x_0) = 0$, we have $F(x_0) = G^{(T-1)}(x_0)$. Therefore

$$x_T = F^{(T)}(x_0) = F^{(T-1)}(G^{(T-1)}(x_0)) = x_0.$$

The sequence cannot have a period less than T. Note that $F(x) < x$ for all x in $(1, \infty)$. Thus $G(x) > x$ for all x. Consequently

$$x_1 = G^{(T-1)}(x_0) > x_2 = G^{(T-2)}(x_0) > \cdots > x_{T-1} = G(x_0) > x_0$$

and x_T is the first term of the sequence to be x_0.

Problem 23. *Newton's sequence* takes the form of a difference equation

$$x_{t+1} = x_t - \frac{f(x_t)}{f'(x_t)}$$

where $t = 0, 1, 2, \ldots$ and x_0 is the initial value at $t = 0$. Let $f : \mathbb{R} \to \mathbb{R}$ be given by

$$f(x) = x^2 - 1$$

and $x_0 \neq 0$.
(i) Find the fixed points of f.
(ii) Find the fixed points of the difference equation.
(iii) Find the exact solution of

$$x_{t+1} = \frac{1}{2}\left(x_t + \frac{1}{x_t}\right).$$

(iv) Let $x_0 = 1/2$. Find x_1 and x_2.

Solution 23. (i) From the quadratic equation $x^{*2} - 1 = x^*$ we obtain $x_1^* = 1/2 + \sqrt{5}/2$ and $x_2^* = 1/2 - \sqrt{5}/2$, i.e. the *golden mean numbers*.
(ii) Inserting f into the difference equation yields

$$x_{t+1} = \frac{1}{2}\left(x_t + \frac{1}{x_t}\right).$$

Thus the fixed points are the solutions of the equation

$$x^* = \frac{1}{2}\left(x^* + \frac{1}{x^*}\right).$$

We obtain the two fixed points $x^* = \pm 1$.
(iii) The exact solution can be expressed analytically using the *hyperbolic cotangent* and its inverse

$$x_t = \coth(2^t \operatorname{arccoth}(x_0))$$

since the *hyperbolic cotangent* has the identity

$$\coth(2\alpha) \equiv \frac{1+\coth^2(\alpha)}{2\coth(\alpha)}.$$

Note that $\text{arccoth}(0) = \pi/2$. The solution also holds if x_t is complex. The complex plane is divided into two basins of attraction along the imaginary axis, while the imaginary axis in this case is the *Julia set*. The dynamics on it is chaotic.
(iv) With $x_0 = 1/2$ we obtain $x_1 = 5/4$ and $x_2 = 41/40$. The series tends to the fixed point $x^* = 1$.

Problem 24. Let $f : \mathbb{R} \to \mathbb{R}^+$ be a positive, continuously differentiable function, defined for all real numbers and whose derivative is always negative. Show that for any real number x_0 (initial value) the sequence x_t obtained by *Newton's method*

$$x_{t+1} = x_t - \frac{f(x_t)}{f'(x_t)} \qquad t = 0, 1, 2, \ldots$$

has always limit ∞.

Solution 24. Since the derivative $f'(x)$ is always negative while the function $f(x)$ is always positive, the sequence x_t is increasing. Therefore, if the sequence did not have limit ∞, it would be bounded and has a finite limit L. Taking the limit of both sides of the equation and using that f is continuously differentiable yields

$$L = L - \frac{f(L)}{f'(L)}$$

and thus $f(L) = 0$. This is a contradiction.

Problem 25. Let $g : \mathbb{R} \to \mathbb{R}$ be a nonconstant analytic function. Let $\widetilde{x} \in \mathbb{R}$ such that $g(\widetilde{x}) = 0$. Show that if $dg(x = \widetilde{x})/dx \neq 0$, then $\widetilde{x}$ is a *superstable fixed point* for Newton's method with

$$f(x) := x - \frac{g(x)}{dg/dx}.$$

Give an example.

Solution 25. We obtain for the derivative of f

$$\frac{df(x)}{dx} = \frac{g(x)d^2g/dx^2}{(dg/dx)^2}.$$

Since $g(\widetilde{x}) = 0$, $dg/dx(x = \widetilde{x})/dx \neq 0$ we have $df(x = \widetilde{x})/dx = 0$. An example is $g(x) = \sin(x)$.

Problem 26. Consider the analytic function $f : [0,1] \to [0,1]$

$$f(x) = \frac{1}{2} - \frac{1}{2}\sin(2\pi x).$$

Find the fixed points and study their stability.

Solution 26. The fixed points are given by the solution of the equation

$$\frac{1}{2} - \frac{1}{2}\sin(2\pi x^*) = x^*.$$

Since $\sin(\pi) = 0$ one fixed point is given by $x_1^* = 1/2$. The two other fixed points which lie in the intervals $[0, 1/4]$ and $[3/4, 1]$ respectively, must be found numerically for example with the Newton method. In R the following lines

```
uniroot(function(x)
0.5-x-0.5*sin(2.0*pi*x),lower=0.0,upper=0.25,tol=1e-10)
uniroot(function(x)
0.5-x-0.5*sin(2.0*pi*x),lower=0.75,upper=1.0,tol=1e-10)
```

will do the job. We obtain

$$x_2^* = 0.1317579, \quad x_3^* = 0.8682422.$$

To study the stability of the fixed point we start of with

$$\frac{df(x)}{dx} = -\pi\cos(2\pi x).$$

Thus $|df(x = x_1^*)/dx| = \pi > 1$. Hence the fixed point is unstable. Analogously we find for the two other fixed points that $|df(x = x_2^*)/dx| > 1$ and $|df(x = x_3^*)/dx| > 1$, i.e. these fixed points are also unstable.

Problem 27. Let $x \geq 0$ and $r > 0$. Consider the analytic map

$$f_r(x) = xe^{r-x}.$$

(i) Find the fixed points. Study the stability of the fixed points.
(ii) Show that f has a least one periodic point x^* with $x^* \neq 0$ or r.

Solution 27. (i) We find $f(0) = 0$, $f(r) = r$. Thus we have two fixed points $x_1^* = 0$ and $x_2^* = r$. Since

$$\frac{df_r(x)}{dx} = e^{r-x} - xe^{r-x}$$

we find that the fixed point $x_1^* = 0$ is unstable. For the fixed point $x_2^* = r$ we find $f'(r) = 1 - r$. So the stability depends on the parameter r.
(ii) We have for the second iterate

$$f^{(2)}(x) \equiv f(f(x)) = x\exp(r-x)\exp(r - x\exp(r-x)).$$

Thus

$$\left.\frac{d}{dx}f^{(2)}(x)\right|_{x=0} = e^{2r} > 1$$

and

$$\left.\frac{d}{dx}f^{(2)}(x)\right|_{x=r} = (1-r)^2 > 0.$$

For $\epsilon > 0$ sufficiently small we can write

$$f^{(2)}(0+\epsilon) \approx f^{(2)}(0) + \epsilon \left.\frac{d}{dx}f^{(2)}(x)\right|_{x=0} > \epsilon.$$

Thus $f^{(2)}(\epsilon) > \epsilon$. We set $a = \epsilon$. By a similar argument there exists a $b < r$ such that $f^{(2)}(b) < b$. Let $g(x) = f^{(2)}(x) - x$. Therefore, $g(a) > 0$ and $g(b) < 0$. By the intermediate value theorem, there is a x^* so that $g(x^*) = 0$. Thus $f^{(2)}(x^*) = x^*$.

Problem 28. Consider the map $f : \mathbb{R} \to \mathbb{R}$

$$f(x) = -\frac{1}{2}x^2 - x + \frac{1}{2}.$$

Then $f(1) = -1$, $f(-1) = 1$. Thus we have a periodic orbit We set $x_1 = 1$, $x_2 = -1$. Is the orbit attracting or repelling? We have to test

$$|(df(x=x_1)/dx)(df(x=x_2)/dx)| < 1$$

for attracting and

$$|df(x=x_1)/dx)(df(x=x_2)/dx)| > 1$$

for repelling.

Solution 28. Since $df(x)/dx = -x-1$ we have $df(x=x_1)/dx = -2$ and $df(x=x_2)/dx = 0$. Thus the orbit is attracting.

Problem 29. Consider the logistic maps $f_r : \mathbb{R} \to \mathbb{R}$ given by

$$f_r(x) = rx(1-x), \quad r > 0.$$

(i) Find the fixed points of f_r.

(ii) Find for which values of the bifurcation parameter r the fixed points of f_r are attractive.
(iii) Find the periodic points of prime period 2 for f_r. In both cases, establish for which values of r the points will occur.
(iv) Find for which values the periodic points of prime period 2 are attractive.

Solution 29. (i) The fixed points x^* satisfy the equation

$$f_r(x^*) = rx^*(1 - x^*) = x^*.$$

The map therefore has two fixed points, namely

$$x_1^* = 0, \qquad x_2^* = \frac{r-1}{r}.$$

These two points exist for all $r \neq 0$ and, with the exception of one special case ($r = 1$), are always distinct.
(ii) The first derivative of the logistic map is given by

$$\frac{df_r(x)}{dx} = r(1 - 2x).$$

Inserting the fixed points $x_1^* = 0$ and $x_2^* = (r-1)/r$ yields

$$f_r'(x_1^*) = r, \qquad f_r'(x_2^*) = 2 - r.$$

Therefore x_1^* is attractive for $|r| < 1$, i.e. $0 < r < 1$ and x_2^* is attractive for $|2 - r| < 1$, i.e. $1 < r < 3$.
(iii) Periodic points x of period n satisfy the quartic equation

$$f_r^{(2)}(x) = r^2 x(1 - x)(1 - rx(1 - x)) = x.$$

It follows that

$$r^3x^4 - 2r^3x^3 + r^2(r^2 + 1)x^2 + (1 - r^2)x = 0.$$

The fixed points in (i) trivially are also period 2 points. Therefore the expressions x and $x - \frac{r-1}{r}$ can be factored out from the left-hand side yielding the quadratic equation

$$r^2x^2 + r(r + 1)x + r + 1 = 0.$$

Hence the prime-period 2 points of the map f_r are given by

$$x_{3,4} = \frac{1}{2r}\left(r + 1 \pm \sqrt{(r + 1)(r - 3)}\right).$$

These points only exist for $r \geq 3$.

(iv) The periodic points of f_r of prime period 2 are

$$x_{3,4} = \frac{1}{2r}\left(r + 1 \pm \sqrt{(r+1)(r-3)}\right).$$

Applying the chain rule for the derivative of the second iterate of f_r and substituting the result one finds

$$(f^{(2)})'(x) = f_r'(f_r(x))\, f_r'(x) = r[1 - 2f_r(x)]\, r(1 - 2x).$$

Simplifying provides

$$(f^{(2)})'(x_3) = r^2[1 - 2f_r(x_3)](1 - 2x_3) = r^2(1 - 2x_4)(1 - 2x_3) = -r^2 + 2r + 4.$$

The period 2 cycle $\{x_3, x_4\}$ therefore is attractive for $|-r^2 + 2r + 4| < 1$ i.e. $3 < r < 1 + \sqrt{6}$. The upper limit of this range, $1 + \sqrt{6} = 3.4495\ldots$, is obviously the value of r for which a second period-doubling bifurcation occurs.

Problem 30. The logistic family $f_r : \mathbb{R} \to \mathbb{R}$ is defined by

$$f_r(x) = rx(1 - x)$$

(i) Show that f_r undergoes a period-doubling bifurcation.
(ii) Find the values of x and r where period doubling occurs.

For f_r to undergo a period-doubling bifurcation for $x = x_0$, $r = r_0$, it must satisfy the following four conditions.

(a) $f_{r_0}(x_0) = x_0$
(b) $f'_{r_0}(x_0) = -1$
(c) $\left.\dfrac{\partial (f_r^{(2)})'(x_0)}{\partial r}\right|_{r=r_0} \neq 0$
(d) $f'''_{r_0}(x_0) \neq -\frac{3}{2}(f''_r(x_0))^2$.

Solution 30. (i) The conditions (a) and (b) determine x_0 and r_0. The derivative of f_r is given by

$$\frac{df_r(x)}{dx} = r(1 - 2x).$$

The system of equations

$$r_0 x_0(1 - x_0) = x_0, \quad r_0(1 - 2x_0) = -1$$

results with the unknowns r_0 and x_0. The solution is $x_0 = \frac{2}{3}$, $r_0 = 3$. To establish whether condition (c) is satisfied, consider

$$(f^{(2)})'(x) = f_r'(f_r(x))\, f_r'(x)$$

where the chain rule has been applied. Substituting f_r' we obtain

$$(f^{(2)})'(x) = r[1-2f_r(x)]\,r(1-2x) = r^2(1-2x) - r^3x(1-x)(1-2x).$$

Therefore

$$\frac{\partial (f_r^{(2)})'(x)}{\partial r} = 2r(1-2x) - 3r^2x(1-x)(1-2x).$$

With $r = r_0$ and $x = x_0$ we find

$$\left.\frac{\partial (f_r^{(2)})'(x_0)}{\partial r}\right|_{r=r_0} = -1 \neq 0$$

and condition (c) is satisfied. To establish whether condition (d) is satisfied, we utilize

$$f_r''(x) = -2r$$

and $f_r'''(x) = 0$. If the values are substituted it follows that

$$f_{r_0}'''(x_0) = 0 \neq -\frac{3}{2} = -\frac{3}{2}(f_r''(x_0))^2$$

and condition (d) is satisfied. A period-doubling bifurcation therefore occurs for the values $x_0 = 2/3$ and $r_0 = 3$.

Problem 31. The family of quadratic maps $f_r : \mathbb{R} \to \mathbb{R}$ is defined by

$$f_r(x) = x^2 + r.$$

Find out whether f_r undergoes a *tangent bifurcation*, if so, for which values of x and r it occurs.
For f_r to undergo a tangent bifurcation for $x = x_0$, $r = r_0$, it must satisfy the following four conditions.

(a) $f_{r_0}(x_0) = x_0$
(b) $f_{r_0}'(x_0) = 1$
(c) $f_{r_0}''(x_0) \neq 0$
(d) $\left.\dfrac{\partial f_r(x_0)}{\partial r}\right|_{r=r_0} \neq 0$

Solution 31. The conditions (a) and (b) determine x_0 and r_0. The derivative of f_r is given by $f_r'(x) = 2x$. Therefore the two simultaneous equations result

$$x_0^2 + r_0 = x_0, \qquad 2x_0 = 1.$$

The solution is $x_0 = \frac{1}{2}$, $r_0 = \frac{1}{4}$. The conditions (c) and (d) are satisfied

$$f''_{r_0}(x_0) = 2x_0 \neq 0, \qquad \left.\frac{\partial f_r(x_0)}{\partial r}\right|_{r=r_0} = 1 \neq 0.$$

A tangent bifurcation therefore occurs for the values $x_0 = 1/2$, $r_0 = 1/4$.

Problem 32. Consider the logistic family $f_r : \mathbb{R} \to \mathbb{R}$ given by

$$f_r(x) = rx(1-x).$$

Show that there exists an infinite number of eventually fixed points for $r > 4$.

Solution 32. For all values of the parameter the logistic map has fixed points at 0 and $\frac{r-1}{r}$. For $r > 4$, both these points lie in the unit interval $[0,1]$. Consider $x \in [0,1]$. Let y be a *preimage* of a given x under the map

$$f_r(x) = rx(1-x) \qquad r > 4$$

i.e. $f_r(y) = x$. We obtain the quadratic equation $ry^2 - ry + x = 0$ with the two solutions

$$y_{1,2} = \frac{1}{2}\left(1 \pm \sqrt{1 - \frac{4x}{r}}\right).$$

Since $0 \leq x \leq 1$ and $r > 4$ it then follows that $0 \leq 4x/r < 1$. Thus both the preimages exist. Furthermore one has

$$0 < \sqrt{1 - \frac{4x}{r}} \leq 1.$$

Substituting this inequality one finds that

$$\frac{1}{2} < y_1 \leq 1 \text{ and } 0 \leq y_2 < \frac{1}{2}.$$

Hence $y_1, y_2 \in [0,1]$. It follows that $x \in [0,1]$ always has two distinct preimages on $[0,1]$. The two fixed points each have two preimages: from (1), the fixed point at 0 has the preimages 0 (trivially) and 1 while the fixed point at $\frac{r-1}{r}$ has preimages at $\frac{r-1}{r}$ (trivially) and at $\frac{1}{r}$. One finds that each of the two fixed points have 2^n preimages of order n and hence $2(2^n - 1)$ preimages of order up to n. All of these preimages are therefore eventually fixed at one of the two fixed points. All of these points are distinct.
(a) If two preimages of order n are identical, say x_1, it implies that $m < n$ and $x_2 \neq x_3$ exist such that $f_r^{(m)}(x_1) = x_2$ and $f_r^{(m)}(x_1) = x_3$. This cannot be since f_r is a single-valued function.

(b) Assume that two preimages of the fixed point x_0 of order n and m respectively ($n \neq m$) are identical, say x_1. One can assume that $m < n$. It follows that

$$f_r^{(n-m)}(x_1) = x_1$$

and

$$f_r^{(m)}(x_1) = x_0.$$

Choose k such that $k(n-m) > m$. Then we have

$$\begin{aligned} f_r^{(k(n-m))} &= f_r^{((k-1)(n-m))}\left(f^{(n-m)}(x_1)\right) \\ &= f_r^{((k-1)(n-m))}(x_1) \\ &\vdots \\ &= f_r^{(n-m)}(x_1) \\ &= x_1. \end{aligned}$$

It follows that

$$f_r^{(k(n-m))}(x_1) = f_r^{(k(n-m)-m)}\left(f^{(m)}(x_1)\right) = f_r^{(k(n-m)-m)}(x_0) = x_0$$

since x_0 is a fixed point. Hence $x_1 = x_0$, which represents a special case (order 1). Therefore two preimages of different order cannot be identical. Thus for each of the two fixed points there are 2^n distinct preimages. Since $n \in \mathbb{N}$, there are an infinite number of these eventually fixed points.

Problem 33. The symmetric *tent map* $f : [0,1] \to [0,1]$ is defined by

$$f(x) = \begin{cases} 2x & \text{for } x \in [0,1/2] \\ 2-2x & \text{for } x \in [1/2,1]. \end{cases}$$

(i) Sketch the graph of the symmetric tent map f. Draw the line $f(x) = x$ for $x \in [0,1]$ to locate the fixed points. Find the fixed points.
(ii) Sketch the graph for the second iterate $f^{(2)}$ of the symmetric tent map f. Draw the line the locate the fixed points.
(iii) Let $f^{(n)}$ be the n-th iterate. Show that $f^{(n)}$ has 2^n repelling periodic points of period n.
(iv) Show that these periodic points are dense on $[0,1]$.

Solution 33. (i) The tent map is a piecewise linear continuous function as shown in the graph.

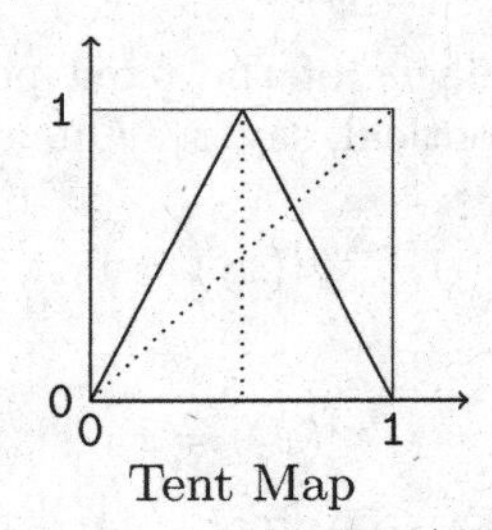

Tent Map

The fixed points are given by 0 and the solution of $2 - 2x^* = x^*$, i.e. $x^* = 2/3$. The two fixed points are unstable
(ii) For its second iterate follows from the definition of the tent map that

$$f^{(2)}(x) = \begin{cases} 2f(x) \text{ for } f(x) \in [0, 1/2] \\ 2 - 2f(x) \text{ for } f(x) \in [1/2, 1]. \end{cases}$$

Consequently

$$f^{(2)}(x) = \begin{cases} 4x \text{ for } x \in [0, 1/4] \\ 2 - 4x \text{ for } x \in [1/4, 1/2] \\ 4x - 2 \text{ for } x \in [1/2, 3/4] \\ 4 - 4x \text{ for } x \in [3/4, 1]. \end{cases}$$

The graph of the second iterate is shown below

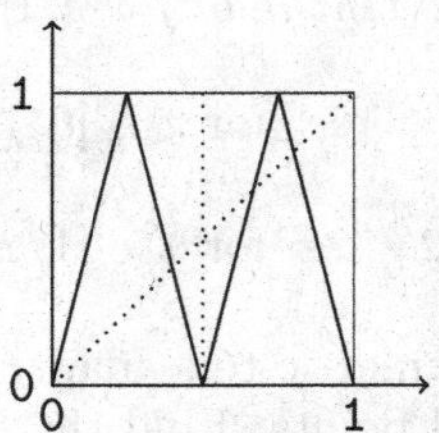

Second Iterate of Tent Map

The four fixed points are given by $x^* = 0$, $x^* = 2/5$ (as solution of $2 - 4x^* = x^*$), $x^* = 2/3$ (as solution of $4x^* - 2 = x^*$) and $x^* = 4/5$ as solution of $4 - 4x^* = x^*$. Obviously the fixed points include the two fixed points of f.
(iii) The equations of two piece-wise linear parts of the k-th period of the graph are respectively given by

$$y = f_1(x) := 2^n x - (2k - 2)$$

and

$$y = f_2(x) := -2^n x + 2k.$$

The period-n points of f occur where the graph of $f^{(n)}$ intersect the diagonal $y = x$. From these two equations one finds these points to be given respectively by

$$x_k = \frac{2k-2}{2^n - 1} \qquad k = 1, 2, \ldots, 2^{n-1}$$

and

$$\xi_k = \frac{2k}{2^n + 1} \qquad k = 1, 2, \ldots, 2^{n-1}.$$

Therefore, there are $2 \cdot 2^{n-1} = 2^n$ periodic points of period n. Furthermore, since $|f_1'(x)| = |f_2'(x)| = 2^n > 1$ on the k-th period, all the period-n points are repelling.
(iv) Let $x \in [0, 1]$, $\epsilon \leq 1$ and let

$$N_\epsilon(x) = (x - \frac{\epsilon}{2}, x + \frac{\epsilon}{2})$$

be a neighbourhood of x. It follows that the largest distance between two points of period n is given by

$$d := \xi_1 - x_1 = \frac{2}{2^n + 1}.$$

Any interval of length d therefore contains at least one periodic point. Now choose N such that $\frac{2}{2^N+1} < \epsilon$. Then, for all $n > N$, there is at least one periodic point in $N_\epsilon(x)$. Therefore, the periodic points are dense on $[0, 1]$.

Problem 34. Consider the symmetric *tent map* f on the unit interval $[0, 1]$

$$f(x) = \begin{cases} 2x & \text{if } x \leq 1/2 \\ 2 - 2x & \text{if } x \geq 1/2 \end{cases}$$

The map is chaotic and completely mixing. It also has a unique absolutely continuous invariant measure and cycles of all orders. Let N be a positive integer. As a discrete model g of f consider the restriction of f on the N-digital binary lattice

$$L_N = \left\{ 0, \frac{1}{2^N}, \frac{2}{2^N}, \frac{3}{3^N}, \ldots, \frac{2^N - 1}{2^N} \right\}.$$

Show that g is asymptotically trivial.

Solution 34. We have $g^{(k)} = 0$ if $k \geq N$. The map g has only the zero cycle and only one invariant measure concentrated at 0.

Problem 35. Consider the function $f : [0, \infty) \to [0, \infty)$

$$f(x) = x^4 e^{-x}$$

or written as difference equation

$$x_{t+1} = x_t^4 e^{-x_t}, \quad t = 0, 1, 2, \ldots$$

with $x_0 \geq 0$.
(i) Find the fixed points of f and study their stability.
(ii) Find the maxima and minima of the function f.

Solution 35. (i) From $f(x^*) = x^*$ we obtain the trivial fixed point $x_0^* = 0$. To find other fixed points we have to solve $x^{*3}e^{-x^*} = 1$. This has to be done numerically and we obtain two more fixed points $x_1^* \approx 1.8571...$ and $x_2^* = 4.5364....$ Now the derivative of f is

$$\frac{df(x)}{dx} = x^3 e^{-x}(4 - x).$$

For the fixed point $x_0^* = 0$ we have $df(x = x_0^*)/dx = 0$. Thus this fixed point is stable. For the second fixed point x_1^* we find $df(x = x_1^*)/dx > 1$. Thus this fixed point is unstable. Finally the fixed point x_2^* provides $|df(x = x_2^*)/dx| < 1$ and hence the fixed point is stable. The following Maxima program will do the job

```
r1: find_root(x^3*exp(-x)-1,x,1,2);
r2: find_root(x^3*exp(-x)-1,x,3,5);
f: x*x*x*x*exp(-x);
fd: diff(f,x); /* derivative of f with respect to x */
r3: subst(r1,x,fd);
r4: subst(r2,x,fd);
```

(ii) From $df(x)/dx = 0$ we obtain the critical points $x = 0$ and $x = 4$. Now the second derivative of f is

$$\frac{d^2 f(x)}{dx^2} = (12x^2 - 8x^3 + x^4)e^{-x}.$$

Thus $d^2 f(x = 4)/dx^2 < 0$ and we have a maximum. Obviously for $x = 0$ we have a minimum.

Problem 36. Consider the *Bernoulli map* $f : [0, 1) \to [0, 1)$, $f(x) = 2x \bmod 1$. Let $i, j \in \mathbb{N}$, $0 \leq k < 2^i$. Consider the interval

$$I = [k/2^i, (k + 1)/2^i] \subset [0, 1].$$

Find the *preimage* $f^{(-1)}(I)$ and the preimage $f^{(-n)}(I)$.

Solution 36. Two intervals map into I. We obtain

$$f^{(-1)}(I) = [k/2^{i+1}, (k+1)/2^{i+1}] \cup [k/2^{i+1} + 1/2, (k+1)/2^{i+1} + 1/2].$$

Hence the preimage $f^{(-1)}$ consists of 2 intervals of length $1/2^{i+1}$ spaced by $1/2$. Consequently $f^{(-n)}(I)$ consists of 2^n intervals of length $1/2^{i+n}$ spaced by $1/2^n$.

Problem 37. Let $f : [0,1) \to [0,1)$. The *Bernoulli map* is defined by

$$f(x) := 2x \bmod 1.$$

The map can be written as a difference equation

$$x_{t+1} = \begin{cases} 2x_t & \text{for } 0 \le x_t < 1/2 \\ (2x_t - 1) & \text{for } 1/2 \le x_t < 1 \end{cases}$$

where $t = 0, 1, 2, \ldots$ and $x_0 \in [0,1)$.
(i) Find the fixed points.
(ii) Study the stability of the fixed points.
(iii) Find a periodic orbit.
(iv) Find the exact solution.
(v) Evaluate the Liapunov exponent.
(vi) Find the invariant density.
(vii) Evaluate the autocorrelation function.

Solution 37. (i) The fixed points are defined by $x^* = f(x^*)$. Thus for the Bernoulli map we find only one fixed point $x^* = 0$.
(ii) The fixed point is unstable owing to $df(x)/dx = 2$ in the interval $[0, 1/2]$.
(iii) Let $\mathbb{Q} \cap [0,1]$, where $\mathbb{Q}$ denotes the rational numbers. Then x_t is a periodic orbit or tends to the fixed point $x^* = 0$.
(iv) The exact solution is given by $x_t = 2^t \bmod 1$.
(v) Owing to $df/dx = 2$ in the interval $[0, 1/2]$ and $df/dx = 2$ in the interval $[1/2, 1]$ the Liapunov exponent for almost all initial values is given by $\ln(2)$.
(vi) The invariant density is given by $\rho(x) = 1$.
(vii) The autocorrelation function is defined by

$$C_{xx}(\tau) := \lim_{T\to\infty} \frac{1}{T} \sum_{t=0}^{T} (x_t - \langle x_t \rangle)(x_{t+\tau} - \langle x_t \rangle).$$

For almost all initial value we find

$$C_{xx}(\tau) = \frac{2^{-\tau}}{12}, \quad \tau = 0, 1, 2, \ldots .$$

Problem 38. Consider the first-order discrete time dynamical system

$$x_{t+1} = 2x_t \mod 1, \quad t = 0, 1, 2, \ldots$$

and

$$s_t = \begin{cases} 1 \text{ if } x_t \geq 0.5 \\ 0 \text{ if } x_t < 0.5 \end{cases}$$

where $x_0 \in [0, 1]$. We call $\mathbf{s} = s_0 s_1 s_2 \ldots$ the output symbol. Show that if $x_0 \in [0.78125, 0.8125]$ then the output coincide for the first three bits.

Solution 38. For $x_0 = 0.78125$ we have

$$0.78125 \to 1, \quad 0.56250 \to 1, \quad 0.12510 \to 0, \quad 0.25020 \to 0.$$

For $x_0 = 0.8125$ we have

$$0.8125 \to 1, \quad 0.6250 \to 1, \quad 0.2500 \to 0, \quad 0.5000 \to 1.$$

Problem 39. (i) Let $f : [-1, 1] \mapsto [-1, 1]$ be defined by

$$f(x) := 1 - 2x^2. \tag{1}$$

Let $-1 \leq a \leq b \leq 1$ and

$$\mu([a, b]) := \frac{1}{\pi} \int_a^b \frac{dx}{\sqrt{1 - x^2}}. \tag{2}$$

Calculate $\mu([-1, 1])$.
(ii) Show that

$$\mu(f^{-1}([a, b])) = \mu([a, b]) \tag{3}$$

where $f^{-1}([a, b])$ denotes the set S which is mapped under f to $[a, b]$, i. e. $f(S) = [a, b]$. The quantity μ is called the *invariant measure* of the map f.

Solution 39. (i) Since

$$\int_a^b \frac{dx}{\sqrt{1 - x^2}} = \arcsin(b) - \arcsin(a)$$

and $\arcsin(1) = \pi/2$, $\arcsin(-1) = -\pi/2$ we obtain $\mu([-1, 1]) = 1$.
(ii) The inverse function f^{-1} is not globally defined. We set $f_1 : [-1, 0] \mapsto [-1, 1]$ with $f_1(x) = 1 - 2x^2$. Then $f_1^{-1} : [-1, 1] \mapsto [-1, 0]$,

$$f_1^{-1}(x) = -\sqrt{\frac{1 - x}{2}}$$

with $f_1^{-1}(-1) = -1$ and $f_1^{-1}(1) = 0$. Analogously, we set $f_2 : [0,1] \mapsto [-1,1]$ with $f_2(x) = 1 - 2x^2$. Then $f_2^{-1} : [-1,1] \mapsto [0,1]$,

$$f_2^{-1}(x) = \sqrt{\frac{1-x}{2}}$$

with $f_2^{-1}(-1) = 1$ and $f_2^{-1}(1) = 0$. It follows that

$$f_1^{-1}(a) = -\sqrt{\frac{1-a}{2}} \qquad f_1^{-1}(b) = -\sqrt{\frac{1-b}{2}}$$

$$f_2^{-1}(a) = \sqrt{\frac{1-a}{2}} \qquad f_2^{-1}(b) = \sqrt{\frac{1-b}{2}}.$$

Thus the mapping f maps the intervals

$$\left[-\sqrt{\frac{1-a}{2}}, -\sqrt{\frac{1-b}{2}}\right], \qquad \left[\sqrt{\frac{1-b}{2}}, \sqrt{\frac{1-a}{2}}\right]$$

into the interval $[a,b]$. From equations (2) we notice that

$$\pi\mu([a,b]) = \arcsin(a) - \arcsin(b).$$

Furthermore $\arcsin(-x) \equiv -\arcsin(x)$. Now

$$\int_{-\sqrt{(1-a)/2}}^{-\sqrt{(1-b)/2}} \frac{dx}{\sqrt{1-x^2}} + \int_{\sqrt{(1-b)/2}}^{\sqrt{(1-a)/2}} \frac{dx}{\sqrt{1-x^2}} = \arcsin(b) - \arcsin(a).$$

Condition (3) can also be written as

$$\frac{1}{\pi}\frac{1}{\sqrt{1-x^2}} = \frac{d}{dx}\int_{f^{-1}([-1,x])} \frac{ds}{\pi\sqrt{1-s^2}}.$$

Problem 40. The *Chebyshev polynomials* of first kind can be defined as $(n = 0, 1, 2, \ldots)$

$$\begin{aligned} T_n(x) &= \frac{1}{2}\left(\left(x - \sqrt{x^2-1}\right)^n + \left(x + \sqrt{x^2-1}\right)^n\right) \\ &= x^n \sum_{k=0}^{\lfloor \frac{n}{2} \rfloor} \binom{n}{2k} (1 - x^{-2})^k \end{aligned}$$

or as the recurrence relation $(n = 1, 2, \ldots)$

$$T_{n+1}(x) = 2xT_n(x) - T_{n-1}(x)$$

with $T_0(x) = 1$, $T_1(x) = x$.
(i) Find T_2 and T_3.
(ii) Find $T_n(T_m(x))$ with $n, m \in \mathbb{N}_0$.
(iii) Find the fixed points of T_3 and study whether they are stable.

Solution 40. (i) Applying the recurrence relation we obtain

$$T_2(x) = 2x^2 - 1, \qquad T_3(x) = 4x^3 - 3x.$$

(ii) We obtain

$$T_n(T_m(x)) = T_{nm}(x).$$

(iii) The fixed point for $T_3(x) = 4x^3 - 3x$ are $x^* = 0$, $x^* = -1$, $x^* = +1$. They are unstable. For $x \in [-1, 1]$ we have $T_3(x) \in [-1, 1]$. For $T_3 : [-1, 1] \to [-1, 1]$ we have a fully developed chaotic map with Liapunov exponent $\ln(3)$.

Problem 41. Consider the *cubic map* $f_r : [-1, 1] \to [-1, 1]$

$$f_r(x) = rx^3 + (1 - r)x, \quad r \in [3.2, 4.0]$$

(i) Find the fixed points and study their stability.
(ii) Find the critical points of the map f_r in the interval $[-1, 1]$ and test whether we have a minimum or maximum.
(iii) Find the linearized map (variational equation)
(iv) Find the exact solution for the case $r = 4$.

Solution 41. (i) The fixed points follow from the solution of the equation

$$rx^{*3} + (1 - r)x^* = x^*$$

as $x^* = 0$, $x^* = -1$, $x^* = 1$. They are unstable for $r \in [3.2, 4.0]$.
(ii) The first and second derivative of f_r are

$$\frac{df_r(x)}{dx} = 3rx^2 + (1 - r), \qquad \frac{d^2 f_r(x)}{dx^2} = 6rx.$$

From $3rx^2 + (1 - r) = 0$ the critical points follows as

$$x_1 = -\sqrt{\frac{r-1}{3r}}, \quad x_2 = \sqrt{\frac{r-1}{3r}}.$$

Looking at the second order derivative the critical point x_1 is a maximum and the critical point x_2 is a minimum. Thus for a *symbolic dynamics* we divide the interval $[-1, 1]$ into the three subintervals

$$-1 \le x < -\sqrt{\frac{r-1}{3r}}, \quad -\sqrt{\frac{r-1}{3r}} \le x < \sqrt{\frac{r-1}{3r}}, \quad \sqrt{\frac{r-1}{3r}} \le x \le 1.$$

(iii) We write the map f_r as difference equation

$$x_{t+1} = rx_t^3 + (1-r)x_t, \quad t = 0, 1, 2, \ldots$$

Then the linearized map is given by

$$y_t = (r(3x_t^2 - 1) + 1)y_t, \quad t = 0, 1, 2, \ldots$$

(iv) For the case $r = 4$ we have $x_{t+1} = 4x_t^3 - 3x_t$. Utilizing the identity

$$\cos(3\alpha) \equiv 4\cos^3(\alpha) - 3\cos(\alpha)$$

we obtain the exact solution

$$x_t = \cos(3^t \arccos(x_0)), \quad t = 0, 1, 2, \ldots$$

with $x_0 \in [-1, 1]$.

Problem 42. Consider the piecewise linear map $f : [0,1] \to [0,1]$

$$f(x) = \begin{cases} 3x & \text{for } 0 \le x \le 1/3 \\ 2 - 3x & \text{for } 1/3 \le x \le 2/3 \\ 3x - 2 & \text{for } 2/3 \le x \le 1. \end{cases}$$

(i) Find the fixed points and study their stability.
(ii) Find $f(1/4)$ and $f(f(1/4))$.

Solution 42. (i) The fixed points are given by solutions of the equations

$$3x^* = x^*, \quad 2 - 3x^* = x^*, \quad 3x^* - 2 = x^*$$

with the solutions for the three fixed points $x_1^* = 0$, $x_2^* = 1/2$, $x^* = 1$. With $f_1(x) = 3x$, $f_2(x) = 2 - 3x$, $f_3(x) = 3x - 2$ and $df_1/dx = 3$, $df_2/dx = -3$, $df_3/dx = 3$ we find that all three fixed points are unstable.
(ii) We obtain

$$1/4, \quad f(1/4) = 3/4, \quad f(3/4) = 1/4.$$

Thus we have a periodic orbit.

Problem 43. Consider the piecewise linear map $f : [0,1] \to [0,1]$

$$f(x) = \begin{cases} 4x & \text{for } 0 \le x < 1/4 \\ 4x - 1 & \text{for } 1/4 \le x < 1/2 \\ -4x + 3 & \text{for } 1/2 \le x < 3/4 \\ -4x + 4 & \text{for } 3/4 \le x \le 1. \end{cases}$$

(i) Find the fixed points and study their stability.

(ii) Find $f(1/2)$, $f(f(1/2))$, $f(f(f(1/2)))$.
(iii) Find $f(2/3)$ and $f(f(2/3))$.

Solution 43. (i) The fixed points are given by the solutions of the four equations

$$4x^* = x^*, \quad 4x^* - 1 = x^*, \quad -4x^* + 3 = x^*, \quad -4x^* + 4 = x^*$$

with the solutions for the four fixed points $x_1^* = 0$, $x_2^* = 1/3$, $x_3^* = 3/5$, $x_4^* = 4/5$. With $f_1(x) = 4x$, $f_2(x) = 4x - 1$, $f_3(x) = -4x + 3$, $f_4(x) = -4x + 4$ we obtain $df_1/dx = 4$, $df_2/dx = 4$, $df_3/dx = -4$, $df_4/dx = -4$ we find that the fixed points are unstable.
(ii) We obtain

$$1/2, \quad f(1/2) = 1, \quad f(1) = 0$$

i.e. the orbit tends to the fixed point $x^* = 0$.
(iii) We obtain

$$2/3, \quad f(2/3) = 1/3, \quad f(1/3) = 1/3$$

i.e. the orbit tends to the fixed point $x^* = 1/3$.

Problem 44. Consider the symmetric *tent map* given by

$$f(x) = \begin{cases} 2x & \text{for } 0 \le x \le 1/2 \\ 2(1-x) & \text{for } 1/2 \le x \le 1. \end{cases} \tag{1}$$

The *Frobenius-Perron integral equation* is given by

$$\rho(x) = \int_0^1 \rho(y)\delta(x - f(y))dy \tag{2}$$

where δ is the delta function. Find ρ for the symmetric tent map.

Solution 44. Inserting (1) into (2) yields

$$\rho(x) = \frac{1}{2}\left(\rho\left(\frac{x}{2}\right) + \rho\left(1 - \frac{x}{2}\right)\right) \tag{3}$$

where we used the properties of the delta function. Equation (3) has the normalized solution $\rho(x) = 1$. This solution is unique as we can see starting from any arbitrary initial condition $\rho_0(x)$ and iterating n times with (1)

$$\rho_n(x) = \frac{1}{2^n}\sum_{j=1}^{2^{n-1}}\left(\rho_0\left(\frac{j-1}{2^{n-1}} + \frac{x}{2^n}\right) + \rho_0\left(\frac{j}{2^{n-1}} - \frac{x}{2^n}\right)\right)$$

which converges towards

$$\rho(x) = \lim_{n\to\infty} \rho_n(x) = \frac{1}{2}\left(\int_0^1 \rho_0(x)dx + \int_0^1 \rho_0(x)dx\right) = 1.$$

Problem 45. Let $r \geq 1/2$. Consider the map $f : [-1,1] \to [-1,1]$ with

$$f(x) = 1 - 2|x|^r. \tag{1}$$

This map is a fully developed chaotic map for $r \geq \frac{1}{2}$. The *Frobenius-Perron integral equation*

$$\rho(x) = \int_0^1 \rho(y)\delta(x - f(y))dy \tag{2}$$

becomes

$$\rho(x) = \frac{1}{2r}\left(\frac{1-x}{2}\right)^{(1-r)/r}\left[\rho\left(\left(\frac{1-x}{2}\right)^{1/r}\right) + \rho\left(-\left(\frac{1-x}{2}\right)^{1/r}\right)\right]. \tag{3}$$

Find the invariant density for $r = 1$, $r = 1/2$ and $r = 2$.

Solution 45. Consider $r = 1$. This is the symmetric tent map. For $r = 1$ the Frobenius-Perron equation reads

$$\rho(x) = \frac{1}{2}\left(\rho\left(\frac{1-x}{2}\right) + \rho\left(\frac{x-1}{2}\right)\right). \tag{4}$$

With

$$\rho_0(x) = \frac{1}{2}$$

we find

$$\rho_k(x) = \frac{1}{2}, \qquad k = 1, 2, 3, \ldots.$$

Therefore

$$\rho(x) = \frac{1}{2}$$

is the ergodic solution of (4). Consider now $r = 1/2$ (square-root map). Then we have

$$\rho(x) = \frac{1}{2}(1-x)\left[\rho\left(\left(\frac{1-x}{2}\right)^2\right) + \rho\left(-\left(\frac{1-x}{2}\right)^2\right)\right].$$

Starting with $\rho_0(x) = 1/2$ we find

$$\rho_k(x) = \frac{1}{2}(1-x) \qquad k = 1, 2, 3, \ldots.$$

The ergodic solution is therefore

$$\rho(x) = \frac{1}{2}(1-x).$$

Consider now $r = 2$ (quadratic map). Then equation (3) takes the form

$$\rho(x) = \frac{1}{4}\left(\frac{1-x}{2}\right)^{-1/2}\left[\rho\left(\left(\frac{1-x}{2}\right)^{1/2}\right) + \rho\left(-\left(\frac{1-x}{2}\right)^{1/2}\right)\right].$$

Taking $\rho_0(x) = 1/2$ we find

$$\rho_1(x) = \frac{\sqrt{2}}{4}(1-x)^{-1/2} = \frac{1}{2}\frac{\cos(\frac{1}{2}\arccos(x))}{\sqrt{1-x^2}}$$

with $\arccos(x) \in |0, \pi|$. For $\rho_2(x)$ one obtains

$$\rho_2(x) = \frac{1}{2}\cos\left(\frac{1}{4}\pi\right)\frac{\cos(\frac{1}{4}\arccos(x))}{\sqrt{1-x^2}}$$

and in general

$$\rho_k(x) = \frac{1}{2}\left(\prod_{i=2}^{k}\cos\left(\frac{\pi}{2^i}\right)\right)\frac{\cos(\arccos(x)/2^k)}{\sqrt{1-x^2}}, \qquad k = 2, 3, 4, \ldots.$$

Taking the limit $k \to \infty$ and using the identity

$$\prod_{i=2}^{\infty}\cos\left(\frac{\pi}{2^i}\right) \equiv \frac{2}{\pi}$$

we arrive at

$$\rho(x) = \lim_{k\to\infty}\rho_k(x) = \frac{1}{\pi\sqrt{1-x^2}}.$$

Problem 46. Let Σ be the set of all infinite sequences of 0's and 1's. This set is called the sequence space of 0 and 1 or the symbol space of 0 and 1. More precisely

$$\Sigma := \{\,(s_0 s_1 s_2 \ldots) \,:\, s_i = 0 \,\text{or}\, 1\} \tag{1}$$

Let $s = s_0 s_1 s_2 \ldots$ and $t = t_0 t_1 t_2 \ldots$ be elements in Σ. We denote the distance between s and t as $d[s,t]$ and define it by

$$d[s,t] := \sum_{i=0}^{\infty}\frac{|s_i - t_i|}{2^i}. \tag{2}$$

Then

$$0 \le d[s,t] \le \sum_{i=0}^{\infty} \frac{1}{2^i} = 2. \qquad (3)$$

Note that $|s_i - t_i|$ is either 0 or 1.
(i) Show that the distance between $s = 0000\ldots$ and $t = 01010101010\ldots$ is 2/3.
(ii) Let s and t be elements of Σ. Show that if the first $n+1$ digits in s and t are identical, then $d[s,t] \le 1/2^n$. Show that if $d[s,t] \le 1/2^n$, then the first n digits in s and t are identical.

Solution 46. (i) We find

$$\begin{aligned}\sum_{i=0}^{\infty} \frac{|s_i - t_i|}{2^i} = \sum_{i=0}^{\infty} \frac{|t_i|}{2^i} &= \frac{0}{2^0} + \frac{1}{2^1} + \frac{0}{2^2} + \frac{1}{2^3} + \frac{0}{2^4} + \frac{1}{2^5} + \cdots \\ &= \frac{1}{2} \sum_{i=0}^{\infty} \frac{1}{4^i} \\ &= \frac{1}{2}\left(\frac{1}{1-\frac{1}{4}}\right) = \frac{2}{3}.\end{aligned}$$

(ii) Let $s = s_0 s_1 s_2 \ldots$ and $t = t_0 t_1 t_2 \ldots$ be sequences in Σ. The first $n+1$ digits of s are $s_0, s_1, s_2, \ldots, s_n$. Thus, s and t agree on the first $n+1$ digits if and only if $s_i = t_i$ for $i \le n$. Suppose that $s_i = t_i$ for all $i \le n$. Then

$$d[s,t] = \sum_{i=0}^{\infty} \frac{|s_i - t_i|}{2^i} = \sum_{i=0}^{n} \frac{0}{2^i} + \sum_{i=n+1}^{\infty} \frac{|s_i - t_i|}{2^i}.$$

Therefore

$$d[s,t] = \frac{1}{2^{n+1}} \sum_{i=0}^{\infty} \frac{|s_{i+n+1} - t_{i+n+1}|}{2^i} \le \frac{1}{2^{n+1}} \sum_{i=0}^{\infty} \frac{1}{2^i} = \frac{1}{2^n}.$$

On the other hand, if there is $j < n$ such that $s_j \ne t_j$, then

$$d[s,t] = \sum_{i=0}^{\infty} \frac{|s_i - t_i|}{2^i} \ge \frac{1}{2^j} > \frac{1}{2^n}.$$

Problem 47. Let Σ be the set of all bi-infinite sequences of the binary symbols $\{0,1\}$, i.e.

$$\Sigma := \{\, \sigma : \sigma : \mathbb{Z} \to \{0,1\} \,\}. \qquad (1)$$

The elements, σ of Σ are called *symbol sequences* and they are defined by specifying $\sigma(n) = \sigma_n \in \{0, 1\}$ for each $n \in \mathbb{Z}$. We write

$$\sigma := \{\sigma_n\}_{n=-\infty}^{\infty} = \{\ldots \sigma_{-2}\sigma_{-1} \cdot \sigma_0\sigma_1\sigma_2 \ldots\} \tag{2}$$

We consider the dynamics of the map $f : \Sigma \to \Sigma$ defined by

$$f(\sigma)_n := \sigma_{n-1} \tag{3}$$

$n \in \mathbb{Z}$. This is known as a *left-shift* on Σ because it corresponds to moving the binary point one symbol to the left. Show that the left shift $f : \Sigma \to \Sigma$ has periodic orbits of all period as well as aperiodic orbits.

Solution 47. A point $\sigma^* \in \Sigma$ is periodic if

$$f^{(q)}(\sigma^*) = \sigma^* \tag{4}$$

$q \in \mathbb{N}$. If q is the least, positive integer for which (4) is satisfied then σ^* is said to be of period q. Equation (4) will be satisfied if an only if

$$\sigma_n^* = \sigma_{n+q}^* \tag{5}$$

for all $n \in \mathbb{Z}$. It is then easy to find periodic points of f with any given period, q. The required sequence, σ^*, is generated by repetition of a block of symbols of length q that is itself not composed of repetitions of any of its sub-blocks. For example, the point

$$\sigma^* = \{\ldots 11010\overline{11110111011010} \cdot 11110 \ldots\} \tag{6}$$

has period-14, while

$$\sigma^* = \{\ldots 10\overline{1011010}\,\overline{1011010} \cdot 101 \ldots\}$$

satisfies $f^{(14)}(\sigma^*) = \sigma^*$ but has period-7 because $f^{(7)}(\sigma^*) = \sigma^*$. The function f has aperiodic orbits. For instance

$$\sigma = \{\ldots \underbrace{1..110}_{n} \ldots \underbrace{11110}_{4}\underbrace{1110}_{3}\underbrace{110}_{2}\underbrace{10}_{1} \cdot 1011011101111 \ldots\},$$

which contains symbol blocks of the type shown for all $n \in \mathbb{N}$, is such that there is no $q \in \mathbb{N}$ such that $f^{(q)}(\sigma) = \sigma$.

Problem 48. Show that the map

$$f(x) = x + r \mod 1 \tag{1}$$

is not ergodic when r is rational. This means $r = k/m$, where $k \in \mathbb{Z}$, $m \in \mathbb{Z} \setminus \{0\}$.

Solution 48. The set of $x_0 \neq 0$ together with its iterates $x_0 + j/m$, where $j = 0, \ldots, m-1$, is invariant under f but does not have measure zero or unity. Each orbit of f contains m points. On the other hand, if λ is irrational then f is ergodic with respect to the Lebesgue measure and its orbits are dense on the circle.

Problem 49. Let $\#fix(f)$ denote the number of fixed points of the one-dimensional map f. For example the number of fixed points of the logistic map $g(x) = 4x(1-x)$ is 2. For $g(g(x))$ we have four fixed points. The number

$$h_{top}(f) := \lim_{n\to\infty} \frac{1}{n} \ln(\#fix(f^{(n)}))$$

is called the *topological entropy* of the map f. Show that the topological entropy is invariant under conjugation.

Solution 49. Let h, g be such that $h \circ g = f \circ h$. Consider

$$g^{(n)}(x) = x \Longleftrightarrow h(x) = h(g^{(n)}(x)) = f^{(n)}(h(x)).$$

The question now is to show that we have $h(g^{(2)}(x)) = f^{(2)}(h(x))$. We have

$$\begin{aligned} g(g(x)) &= h^{-1}(f(f(h(x)))) = h^{-1}(f(h(g(x)))) \\ &= h^{-1}(f(h(g(x)))) \\ &= h^{-1}(f(f(h(x)))). \end{aligned}$$

This provides the desired result.

Problem 50. The *Bernoulli map* $f : [0,1) \to [0,1)$ is given by

$$f(x) = 2x \bmod 1. \tag{1}$$

We consider the time evolution of a probability density $\rho_n(x)$ (a "state") describing an ensemble of trajectories. The *Frobenius-Perron integral equation* is given by

$$\rho(x) = \int_0^1 \rho(y)\delta(x - f(y))dy.$$

Thus the time evolution of a state $\rho(x)$ under f is given by the *Frobenius-Perron operator* U, defined by

$$\rho_{n+1}(x) = U\rho_n(x) := \frac{1}{2}\left(\rho_n\left(\frac{1}{2}x\right) + \rho_n\left(\frac{1}{2}(x+1)\right)\right).$$

Since ρ is a probability density, it should be integrable, we require it to be in the Banach space $L_1(0,1)$ of Lebesgue integrable functions on $[0,1]$.

One restricts ρ to the Hilbert space $L_2(0,1)$ of square-integrable functions (Lebesgue measure). Consider the states $e_{n,s}$ defined by

$$e_{n,s}(x) := \exp(2\pi i 2^n (2s+1)x)$$

where n is a non-negative integer and s is an integer. Any integer ℓ can be written uniquely as $\ell = 2^n(2s+1)$ for integers $n \geq 0$ and $s \in \mathbb{Z}$. If $n = 0$ then $e_{0,s}(x) = \exp(2\pi i(2s+1)x)$. Now U is a one-sided shift operator, i.e.

$$Ue_{n,s}(x) := \begin{cases} e_{n-1,s}(x), & n > 0 \\ 0, & n = 0. \end{cases}$$

Show that the spectrum of U fills the unit disk in the complex plane.

Solution 50. Consider the state $\phi_{z,s}$ defined by

$$\phi_{z,s}(x) := \sum_{n=0}^{\infty} z^n e_{n,s}(x).$$

It is formally an eigenstate of U with complex eigenvalue z. For $|z| < 1$ the series converges absolutely and uniformly and defines a continuous square-integrable function. Since the spectrum is closed, the spectrum of U therefore fills the unit disk $|z| \leq 1$.

Problem 51. Calculate the invariant density of the logistic map $g : [0,1] \to [0,1]$

$$g(x) = 4x(1-x)$$

without using the *Frobenius-Perron approach* and by making use of the fact that for the symmetric tent map $f : [0,1] \to [0,1]$

$$f(x) = \begin{cases} 2x & \text{for } 0 \leq x < 0.5 \\ 2-2x & \text{for } 0.5 \leq x \leq 1 \end{cases}$$

the invariant density is constant.

Solution 51. The symmetric tent map and the logistic map are conjugate, i.e. there exists a homeomorphism h such that

$$h \circ g = f \circ h$$

holds. The homeomorphism h is given by

$$h^{-1} : [0,1] \to [0,1], \qquad h^{-1}(y) = \frac{2}{\pi}\sin^{-1}(\sqrt{y}).$$

We first prove that this is a conjugation. The symmetric tent map and the logistic map are conjugate. The conjugation is provided by

$$h^{-1}(y) : x_t = \frac{2}{\pi} \sin^{-1}(\sqrt{y_t}).$$

We start with the given conjugation and the symmetric tent map and show that this leads to the logistic map. Expressed by means of y

$$\frac{2}{\pi} \sin^{-1}(\sqrt{y_{t+1}}) = 2\frac{2}{\pi} \sin^{-1}(\sqrt{y_t}) \Longleftrightarrow y_{t+1} = (\sin(2\sin^{-1}(\sqrt{y_t})))^2$$

with

$$\theta := \sin^{-1}(\sqrt{y}) \Longrightarrow \{\sin(\theta) = \sqrt{y}, \cos(\theta) = \sqrt{1-y}\} :$$
$$y_{t+1} = (2\sin(\theta)\cos(\theta))^2 = 4(\sqrt{y_t}\sqrt{1-y_t})^2.$$

Thus $y_{t+1} = 4y_t(1-y_t)$. Similarly, we have for the symmetric tent map expressed by means of y

$$\frac{2}{\pi} \sin^{-1}(\sqrt{y_{t+1}}) = 2(1 - \frac{2}{\pi} \sin^{-1}(\sqrt{y_t}))$$

so that

$$y_{t+1} = \sin^2(\pi - 2\sin^{-1}(\sqrt{y_t})) = (2\sin(\theta)\cos(\theta))^2 = (2\sqrt{y_t}\sqrt{1-y_t})^2$$

or $y_{t+1} = 4y_t(1-y_t)$. We used that $\sin^2(2\theta) \equiv (2\sin(\theta)\cos(\theta))^2$ The derivative of the conjugation $h^{-1}(x)$ is then needed. We obtain

$$(\sin^{-1}(\sqrt{x}))' = \frac{1}{2\sqrt{x(1-x)}}.$$

Hence we can evaluate the invariant density of the logistic map as we have

$$\rho_f(y)dy = \rho_g(x)dx \to \rho_f(y) = \rho_g(x)\frac{dx}{dy} = \frac{1}{\pi\sqrt{x(1-x)}}.$$

Problem 52. Consider the *Frobenius-Perron integral equation* given by

$$\rho_{n+1}(x) = \int_0^1 \delta(x - f(y))\rho_n(y)dy, \qquad n = 0, 1, 2, \cdots \tag{1}$$

with the map $f : [0,1] \to [0,1]$. Equation (1) implies that the probability density evolves toward the stationary invariant density ρ, provided that all eigenvalues except unity are located within the unit circle. Therefore, if the Frobenius-Perron integral equation is asymptotically stable, then we have

$$\rho(x) = \int_0^1 \delta(x - f(y))\rho(y)dy. \tag{2}$$

Find f if the invariant density ρ is given, where we assume that $f(0) = 0$. Furthermore we assume that

$$f(1-x) = f(x), \qquad \text{(type 1)} \tag{3}$$

$$f(x+1/2) = f(x), \qquad \text{(type 2)}. \tag{4}$$

In type 1, the map is symmetric about the value $x = 1/2$. Type 2 corresponds to the translationally symmetric map.

Solution 52. We substitute f into (3), i.e. we replace x by f, yields

$$\rho(f(x)) = \int_0^1 \delta(\rho(x) - f(y))\rho(y)dy. \tag{5}$$

The *delta function* δ can be written as

$$\delta(f(x) - f(y)) = \frac{\delta(x-y) + \delta(1-x-y)}{\left|\frac{df(x)}{dx}\right|} \tag{6}$$

in type 1, and

$$\delta(f(x) - f(y)) = \frac{\delta(x-y) + \delta(x+1/2-y)}{\left|\frac{df(x)}{dx}\right|} \tag{7}$$

in type 2. By substituting (6) and (7) into (5), we find that

$$\frac{df(x)}{dx} = \frac{\rho(x) + \rho(1-x)}{\rho(f(x))} \tag{8}$$

in type 1, and

$$\frac{df(x)}{dx} = \frac{\rho(x) + \rho(x+1/2)}{\rho(f(x))} \tag{9}$$

in type 2, where we used $f(0) = 0$. The gradient of f is positive in the interval $[0, 1/2]$ for both types. Equations (8) and (9) are the ordinary differential equations for f. Thus the analytical expressions of f can be derived for the appropriate density ρ. The formulae (8) and (9) can be used for deriving a class of the analytically expressed maps for the analytical invariant density. Equations (8) and (9) must be integrated for the interval $[0, 1/2]$. The behaviour of f for $[1/2, 1]$ can be found from the properties of (4) and (5). We obtain $f(1/2-0) = 1$ for both cases. We derive the Liapunov exponent λ given by

$$\lambda := \int_0^1 \rho(x) \ln\left|\frac{df(x)}{dx}\right| dx.$$

Since the forms of df/dx are obtained from (8) and (9), we find that

$$\lambda = \int_0^1 g(x)\ln(g(x))dx - \int_0^1 dx\rho(x)\ln(\rho(x)) + \ln(2)$$

where

$$g(x) := \frac{\rho(x) + \rho(1-x)}{2}$$

for type 1, and

$$g(x) := \frac{\rho(x) + \rho(x+1/2)}{2}$$

for type 2. Here we have assumed that

$$\rho(x+1) = \rho(x)$$

in type 2. It follows that the Liapunov exponent λ is different from $\ln(2)$ by the difference between the information for $\rho(x)$ and $g(x)$. If $\rho(x) = \rho(1-x)$ or $\rho(x) = \rho(x+1/2)$, then we find that λ equals $\ln(2)$.

Problem 53. Consider a one-dimensional ergodic and chaotic map $f : [0,1] \to [0,1]$. The Frobenius-Perron integral equation is given by

$$\rho(x) = \int_0^1 \delta(x - f(y))\rho(y)dy \tag{1}$$

where ρ is the invariant density, i.e.

$$\int_0^1 \rho(x)dx = 1, \qquad \rho(x) > 0 \quad \text{for} \quad x \in [0,1]. \tag{2}$$

Assume that

$$f(0) = 0, \qquad f(1-x) = f(x), \quad x \in [0,1] \tag{3a}$$

and

$$\frac{df(x)}{dx} > 0, \quad x \in [0,1/2]. \tag{3b}$$

Thus f is symmetric about the value $1/2$. Find f if ρ is given.

Solution 53. From (1) we find that

$$\rho(f(x)) = \int_0^1 \delta(f(x) - f(y))\rho(y)dy. \tag{4}$$

The *delta function* has the property

$$\delta(g(u)) = \sum_n \frac{1}{\left|\frac{dg(u=u_n)}{du}\right|}\delta(u - u_n) \tag{5}$$

where the sum runs over all zero's of the function g. Thus we find that

$$\delta(f(x) - f(y)) = \frac{\delta(x-y) + \delta(1-x-y)}{\left|\frac{df(x)}{dx}\right|}. \tag{6}$$

Inserting (6) into (4) yields the differential equation

$$\frac{df}{dx} = \frac{\rho(x) + \rho(1-x)}{\rho(f(x))} \tag{7}$$

where $x \in [0, 1/2]$ and the initial condition is given by $f(0) = 0$. For example let $\rho(x) = 1$. Then we find $f(x) = 2x$ for $x \in [0, 1/2]$. Thus f is the symmetric tent map. Let

$$\rho(x) = \frac{1}{\pi\sqrt{x(1-x)}}.$$

Then using

$$\int \frac{dx}{\sqrt{x(1-x)}} = \arctan\sqrt{\frac{x}{1-x}}$$

we find from (7) that $f(x) = 4x(1-x)$, i.e. the logistic map.

Problem 54. (i) Consider the continuous map $f : [-1, 1] \to [-1, 1]$

$$f(x) = 1 - 2(|x|)^{1/2}$$

or written as difference equation

$$x_{t+1} = 1 - 2(|x_t|)^{1/2}, \qquad t = 0, 1, \ldots$$

with $x_0 \in [-1, 1]$. Find the fixed points and show that they are unstable.
(ii) Find the invariant density and under the assumption that the system is ergodic. Calculate the Liapunov exponent.

Solution 54. (i) We have $f(-1) = -1$, $f(0) = 1$, $f(1) = -1$. The fixed points are given by $x^* = -1$ and $x^* = 3 - \sqrt{8}$. Both fixed points are unstable.
(ii) The Frobenius-Perron integral equation is given by

$$\rho(x) = \int_{-1}^{+1} dy\delta(x - 1 + 2(|y|)^{1/2})\rho(y).$$

Now

$$\rho(1-2x) = \frac{1}{2}\int_{-1}^{+1} dy\delta(|y|^{1/2} - x)\rho(y) = x\rho(x^2) + x\rho(-x^2)$$

with the solution given by the linear function

$$\rho(x) = \frac{1}{2}(1-x).$$

The one dimensional Liapunov exponent follows as

$$\lambda = \int_{-1}^{+1} \rho(x) \ln \left| \frac{df}{dx} \right| dx.$$

It follows that

$$\lambda = \int_{-1}^{+1} \frac{1}{2}(1-x)\ln(|x|^{1/2})dx = \frac{1}{2}.$$

Thus the map is chaotic.

Problem 55. The *W-map* $f : [0,1] \to [1/4, 3/4]$ is given by

$$f(x) = \begin{cases} -2x + 3/4 & \text{for} \quad 0 \le x < 1/4 \\ 2x - 1/4 & \text{for} \ 1/4 \le x < 1/2 \\ -2x + 7/4 & \text{for} \ 1/2 \le x < 3/4 \\ 2x - 5/4 & \text{for} \quad 3/4 \le x \le 1 \end{cases}$$

The graph of the map looks like a W. Find the invariant measure ρ.

Solution 55. We obtain

$$\rho(x) = \begin{cases} 0 & \text{for} \quad 0 \le x < 1/4 \\ 2 & \text{for} \ 1/4 \le x < 1/2 \\ 2 & \text{for} \ 1/2 \le x < 3/4 \\ 0 & \text{for} \quad 3/4 \le x \le 1 \end{cases}$$

with

$$\int_0^1 \rho(x)dx = 1$$

and

$$\mu(S) = \int_S \rho(x)dx = \int_{f^{(-1)}(S)} \rho(x)dx = \mu(f^{(-1)}(S))$$

where $S \subset [0,1]$ (proper subset). For example, assume that $S = [1/4, 3/8]$. Then

$$f^{(-1)}(S) = [3/16, 5/16] \cup [11/16, 13/16].$$

Consequently we have

$$\mu(S) = \int_S \rho(x)dx = \int_{1/4}^{3/8} 2dx = \frac{1}{4}$$

and

$$\begin{aligned}\mu(f^{(-1)}(S)) &= \int_{f^{-1}(S)} \rho(x)dx \\ &= \int_{3/16}^{4/16} 0dx + \int_{4/16}^{5/16} 2dx + \int_{11/16}^{12/16} 2dx + \int_{12/16}^{13/16} 0dx \\ &= \frac{1}{4}.\end{aligned}$$

Problem 56. Consider the one-dimensional map $f : [0,1] \to [0,1]$

$$f(x) = \begin{cases} 1/2 - 2x & \text{for} \quad 0 \le x \le 1/4 \\ -1/2 + 2x & \text{for} \quad 1/4 \le x \le 3/4 \\ 5/2 - 2x & \text{for} \quad 3/4 \le x \le 1 \end{cases}$$

Find the invariant measure.

Solution 56. The invariant measure is given by

$$\rho(x) = \begin{cases} \epsilon & \text{for } 0 \le x \le 1/2 \\ 2 - \epsilon & \text{for } 1/2 \le x \le 1 \end{cases}$$

is invariant, where $\epsilon \in (0,1)$.

Problem 57. Consider the symmetric *tent map* $f : [0,1] \to [0,1]$

$$f(x) = \begin{cases} 2x & \text{for } 0 \le x \le 1/2 \\ -2x + 2 & \text{for } 1/2 < x \le 1 \end{cases}$$

Show that this map can be modelled by the one-sided 2-shift.

Solution 57. Let P_0 be the closed interval $[0, 1/2]$ and P_1 be the closed interval $[1/2, 1]$. Then $P_0 \cap P_1 = \{1/2\}$. The sets P_0 and P_1 have disjoint interiors. Let $\mathcal{P} := \{P_0, P_1\}$. Let $p \in [0,1]$ be a point so that $f^{(k)}(p) \neq 1/2$ for any k. Then for every k, $f^{(k)}(p)$ lies in either P_0 or P_1. We associate with the point p a sequence $x \in \{0,1\}^{\mathbf{N}}$ by choosing x_j so that $f^{(j)}(p) = P_{x_j}$. This sequence is the $\mathcal{P}$-name of p. The point $1/2 \in [0,1]$ has two $\mathcal{P}$-names, namely

$$(.0100\ldots)$$

and

$$(.1100\ldots).$$

This is because $1/2$ is in both the set P_0 and the set P_1. Now $f(1/2) = 1 \in P_1$ and $f^{(j)} = 0 \in P_0$ for all $j > 1$.

Problem 58. A map $f : [-1,1] \to [-1,1]$ is called *S-unimodal* if it satisfies the following conditions:

(a) $f(0) = 1$.
(b) $f([f(1),1]) = [f(1),1]$.
(c) f is monotonically increasing in $[-1,0]$ and monotonically decreasing in $[0,1]$.
(d) f is at least three times continuously differentiable.
(e) $f''(0) < 0$.
(f) The *Schwarzian derivative* of f is negative, i.e.

$$\frac{f'''(x)}{f'(x)} - \frac{3}{2}\left(\frac{f''(x)}{f'(x)}\right)^2 < 0 \text{ for all } x \neq 0.$$

Does the function

$$g_r(x) = 1 - rx^2, \quad r \in (0,2]$$

satisfy these conditions?

Solution 58. We have $g_r(0) = 1$, $g_r([g_r(1),1]) = [g_r(1),1]$. For $x, x' \in [-1,0]$ we have if $x < x'$, $g_r(x) < g_r(x')$. For $x, x' \in [0,1]$ we have if $x < x'$, $g_r(x) > g_r(x')$. Now

$$\frac{dg_r(x)}{dx} = -2rx, \qquad \frac{d^2g_r(x)}{dx^2} = -2r, \qquad \frac{d^3g_r(x)}{dx^3} = 0.$$

Thus g_r has a negative Schwarzian derivative for all $x \neq 0$. Hence all conditions are satisfied.

Problem 59. The *Schwarzian derivative* of an analytic function f of one real variable is defined by

$$(Sf)(x) := \left(\frac{f''(x)}{f'(x)}\right)' - \frac{1}{2}\left(\frac{f''(x)}{f'(x)}\right)^2 \equiv \frac{f'''(x)}{f'(x)} - \frac{3}{2}\left(\frac{f''(x)}{f'(x)}\right)^2.$$

The Schwarzian derivative can also be written as

$$(Sf)(y) := 6 \lim_{x \to y}\left(\frac{f'(x)f'(y)}{(f(x)-f(y))^2} - \frac{1}{(x-y)^2}\right).$$

(i) Find the Schwarzian derivative of

$$h(x) = \frac{ax+b}{cx+d}, \qquad ad - bc \neq 0.$$

(ii) Let

$$g(x) = \frac{af(x)+b}{cf(x)+d}$$

with $ad - bc \neq 0$. Show that $Sg(x) = Sf(x)$.

Solution 59. (i) We obtain $(Sg)(x) = 0$.
(ii) This is best done using computer algebra. A Maxima program is

```
/* Schwarzian.mac */

depends(f,x);
depends(g,x);
g(x) := (a*f(x)+b)/(c*f(x)+d);
D1: diff(g(x),x);
D2: diff(g(x),x,2);
D3: diff(g(x),x,3);
R: D3/D1 - 3*(D2*D2/(D1*D1))/2;
R: ratsimp(R);
```

where the command `ratsimp(.)` simplifies the rational expression.

Problem 60. Consider the initial value problem of the nonlinear difference equation

$$x_t x_{t+1} + c_1 x_t + c_2 = 0, \qquad t = 0, 1, 2, \ldots \tag{1}$$

where c_1 and c_2 are constant and $x_0 \neq 0$. Show that the substitution

$$x_t = \frac{1}{y_t} + \alpha \tag{2}$$

reduces (1) to the linear difference equation

$$(\alpha + c_1) y_{t+1} + \alpha y_t + 1 = 0 \tag{3}$$

if α satisfies the quadratic equation

$$\alpha^2 + c_1 \alpha + c_2 = 0. \tag{4}$$

Solution 60. From (2) we obtain

$$x_{t+1} = \frac{1}{y_{t+1}} + \alpha. \tag{5}$$

Inserting (2) and (5) into (1) yields

$$\left(\frac{1}{y_t} + \alpha\right)\left(\frac{1}{y_{t+1}} + \alpha\right) + c_1 \left(\frac{1}{y_t} + \alpha\right) + c_2 = 0$$

or

$$\frac{(1 + \alpha y_t)(1 + \alpha y_{t+1}) + c_1 (1 + \alpha y_t) y_{t+1} + c_2 y_{t+1} y_t}{y_t y_{t+1}} = 0.$$

Since $y_{t+1}y_t \neq 0$ we arrive at

$$(\alpha^2 + c_1\alpha + c_2)y_{t+1}y_t + (\alpha + c_1)y_{t+1} + \alpha y_t + 1 = 0.$$

If (4) is satisfied we find (3).

Problem 61. Apply the substitution

$$y_t = \frac{x_t}{x_{t+1}} - b_t, \qquad t = 0, 1, 2, \ldots$$

to the *Riccati equation*

$$y_{t+1}y_t + a_{t+1}y_{t+1} + b_{t+1}y_t = c_{t+1}.$$

Solution 61. We obtain the linear second order difference equation

$$(a_{t+1}b_{t+1} + c_{t+1})x_{t+2} - (a_{t+1} - b_t)x_{t+1} - x_t = 0.$$

Problem 62. Consider the map ($t = 0, 1, 2, \ldots$)

$$x_{t+1} = \frac{x_t + 2}{x_t + 1}, \qquad x_0 \geq 0.$$

(i) Find the fixed points.
(ii) Find

$$\lim_{t \to \infty} x_t.$$

Solution 62. (i) From

$$x^* = \frac{x^* + 2}{x^* + 1}$$

we obtain the two fixed points $x_1^* = \sqrt{2}$ and $x_2^* = -\sqrt{2}$.
(ii) Note that we can write

$$x_{t+1} = \frac{x_t + 2}{x_t + 1} \equiv \frac{x_t + 1 + 1}{x_t + 1} \equiv 1 + \frac{1}{x_t + 1}.$$

Thus we obtain

$$\lim_{t \to \infty} x_t = \sqrt{2}$$

which is a fixed point. For $x_0 = 0$ we obtain $x_1 = 2$, $x_2 = 4/3$. The sequence tends to $\sqrt{2}$.

Problem 63. (i) Show that the difference equation

$$x_{t+1} = x_t(3 - 4x_t^2), \qquad t = 0, 1, 2, \ldots \tag{1}$$

with $x_0 \in [-1, 1]$ admits the three fixed points

$$x^* = 0, \qquad x^* = \pm\frac{1}{\sqrt{2}}. \tag{2}$$

(ii) Show that the solution of the initial value problem is given by

$$x_t = \sin(3^t \arcsin(x_0)). \tag{3}$$

Solution 63. (i) The fixed points are given as the solution of the equation

$$x^* = x^*(3 - 4x^{*2}).$$

Thus we obtain the three fixed points $x^* = 0$ and $x^* = \pm 1/\sqrt{2}$. The three fixed points are unstable.
(ii) From (3) we obtain

$$x_{t+1} = \sin(3^{t+1} \arcsin(x_0)).$$

Owing to the identity $\sin(3x) \equiv 3\sin(x) - 4\sin^3(x)$ we find

$$x_{t+1} = 3\sin(3^t \arcsin(x_0)) - 4\sin^3(3^t \arcsin(x_0)) = 3x_t - 4x_t^3.$$

Problem 64. (i) Consider the difference equation

$$x_{t+1} = 16x_t(1 - x_t)(1 - 2x_t^2)^2, \qquad t = 0, 1, 2, \ldots$$

with $x_0 \in [0, 1]$. Find the fixed points.
(ii) Find the exact solution.

Solution 64. (i) The difference equation admits the four fixed points

$$x^* = 0, \qquad x^* = \frac{3}{4}, \qquad x^* = \frac{1}{8}(5 \pm \sqrt{5}).$$

The fixed points are unstable.
(ii) The solution of the initial value problem is given by

$$x_t = \sin^2(4^t \arcsin(\sqrt{x_0})).$$

Problem 65. (i) Consider the difference equation

$$x_{t+1} = \frac{4x_t(1 - x_t)(1 - k^2 x_t)}{(1 - k^2 x_t^2)^2}, \qquad t = 0, 1, 2, \ldots, \qquad 0 \le k^2 \le 1 \tag{1}$$

with $x_0 \in [0,1]$. Find the fixed points.
(ii) Find the solution of the initial value problem using *Jacobi elliptic functions.* Apply the addition theorems for Jacobi elliptic functions

$$\text{sn}(u \pm v) = \frac{\text{sn}(u)\text{cn}(v)\text{dn}(v) \pm \text{cn}(u)\text{sn}(v)\text{dn}(u)}{1 - k^2\text{sn}^2(u)\text{sn}^2(v)}.$$

Solution 65. (i) The nonlinear difference equation with $x_0 \in [0,1]$ admits the fixed points x^* given by $x^* = 0$ and the solutions of the quartic equation

$$k^4 x^{*4} - 3x^{*2} + 6x^* - 3 = 0.$$

(ii) The solution of the initial value problem in terms of elliptic functions is given by

$$x_t = \text{sn}(2^t \text{sn}^{-1}(\sqrt{x_0}, k), k)$$

where sn(.) denotes the Jacobi elliptic functions.

Problem 66. (i) Find the fixed points of the difference equation

$$x_{t+1} = 16x_t(1 - 2\sqrt{x_t} + x_t), \qquad t = 0,1,2,\ldots \qquad (1)$$

with $x_0 \in [0,1]$.
(ii) Find the solution of (1) of the initial value problem.

Solution 66. (i) With $x_0 \in [0,1]$ we obtain the two fixed points

$$x^* = 0, \qquad x^* = \frac{9}{16}$$

which are unstable.
(ii) The solution of the initial value problem is given by

$$x_t = \sin^4(2^t \arcsin(x_0)^{1/4}).$$

Problem 67. (i) Consider the nonlinear difference equation

$$x_{t+1} = \sqrt{2}x_t(1 - x_t^4)^{1/2}, \qquad t = 0,1,2,\ldots \qquad (1)$$

with $x_0 \in [0,1]$. Find the fixed points.
(ii) Find the solution of the initial value problem.

Solution 67. (i) The nonlinear difference equation admit two fixed points

$$x^* = 0, \qquad x^* = \left(\frac{3}{4}\right)^{1/4}$$

which are unstable.
(ii) The solution of the initial value problem is given by

$$x_t = (\sin(2^t \arcsin(x_0^2)))^{1/2}.$$

Problem 68. (i) Consider the nonlinear difference equation

$$x_{t+1} = (x_t^{2/3} - 1)^3, \qquad t = 0, 1, 2, \ldots$$

with $x_0 \in [-1, 1]$. With $x_0 = -1$ we have $x_1 = 0$, with $x_0 = 0$ we have $x_1 = -1$ and with $x_0 = 1$ we have $x_1 = 0$. Find the fixed points.
(ii) Find the solution of the initial value problem.

Solution 68. (i) The nonlinear difference equation admits the fixed point $x^* = -1/8$. The fixed point is unstable.
(ii) The solution of the initial value problem is given by

$$x_t = \cos^3(2^t \arccos(x_0^2)^{-1/3}).$$

Problem 69. Consider the skew-symmetric tent map $f_r : [0, 1] \to [0, 1]$ given by

$$f(x) = \begin{cases} x/r & \text{for } 0 \leq x \leq r \\ (1-x)/(1-r) & \text{for } r < x \leq 1 \end{cases}$$

where $0.5 \leq r < 1$. Find the probability density ρ. Use the probability density to calculate the Liapunov exponent.

Solution 69. A probability density $\rho(x)$ on $[0, 1]$ is invariant, if for each interval $[c, d] \subset [0, 1]$

$$\int_c^d \rho(x)dx = \int_{f^{-1}([c,d])} \rho(x)dx$$

where

$$f^{-1}([c, d]) := \{\, x : c \leq f(x) \leq d \,\}.$$

Thus we obtain the equation for the probability density

$$\rho(x) = r\rho(rx) - (1 - r)\rho(1 - (1 - r)x)$$

with the solution $\rho(x) = 1$. The Liapunov exponent can be calculated as

$$\lambda = \int_0^1 \ln |f'(x)| \rho(x)dx.$$

Thus

$$\lambda_r = \int_0^r \ln|1/r|dx + \int_r^1 \ln|1/(1-r)|dx = -r\ln(r) - (1-r)\ln(1-r).$$

We could also argue as follows. From $y_j = f_r^{(-1)}(x)$ we obtain $y_1 = rx$ and $y_2 = r + (1-r)x$. This provides the invariant density $\rho(x) = 1$. Applying

$$\lambda_r = \int_0^1 \rho(x)\ln|f'(x)|dx = -r\ln(r) - (1-r)\ln(1-r).$$

For $r = 1/2$ we obtain the maximum value $\lambda = \ln(2)$.

Problem 70. Consider the *decimal map* $f : [0,1) \to [0,1)$ defined by

$$f(x) = 10x \mod 1$$

or

$$f(x) = 10x - j \text{ for } \frac{j}{10} \le x < \frac{j+1}{10}, \quad j = 0, 1, \ldots, 9.$$

(i) Let

$$x = \sum_{k=1}^{\infty} \frac{a_k}{10^k} = .a_1 a_2 \cdots.$$

Find $f(x)$. Find $f^{(n)}(x)$.
(ii) Find the fixed points of f.
(iii) Show that under iteration of f no infinite string of 9's can occur in the expansion of any point.

Solution 70. (i) We have

$$f(x) = \sum_{k=1}^{\infty} \frac{a_{k+1}}{10^k} = .a_2 \cdots a_k \cdots$$

so that $a_1(f(x)) = a_2(x)$. In general we have

$$a_1(f^{(n)}(x)) = a_{n+1}(x).$$

(ii) One fixed point is obviously $x^* = 0$. The other fixed points are

$$0.111\ldots, \quad 0.222\ldots, \quad 0.333\ldots, \quad 0.444\ldots$$

$$0.555\ldots, \quad 0.666\ldots, \quad 0.777\ldots, \quad 0.888\ldots$$

(iii) If we denote the greatest integer not exceeding ζ by $\lfloor\zeta\rfloor$, then we have

$$f(x) = 10x - \lfloor 10x \rfloor = 10x - a_1(x).$$

It follows that

$$\begin{aligned} x &= \frac{a_1(x)}{10} + \frac{f(x)}{10} \\ &= \frac{a_1(x)}{10} + \frac{a_2(x)}{10^2} + \frac{f^{(2)}(x)}{10^2} \\ &\vdots \\ &= \frac{a_1(x)}{10} + \cdots + \frac{a_k(x)}{10^k} + \frac{f^{(k)}(x)}{10^k}. \end{aligned}$$

If $x = \ell/10^k$ (ℓ integer less then 10^k), then $f^{(k)}x = 0$ and

$$x = \frac{a_1(x)}{10} + \cdots + \frac{a_k(x)}{10^k}.$$

If $f^{(k)}(x) \neq 0$ for any $k \geq 0$, taking limits gives

$$x = \frac{a_1(x)}{10} + \cdots + \frac{a_k(x)}{10^k} + \cdots$$

Thus under iteration of f no infinite string's of 9's can occur in the expansion of any point.

Problem 71. Consider the *Bernoulli map*

$$\theta_{t+1} = 2\theta_t \mod 1$$

with initial value $\theta_0 \in [0, 1)$.
(i) Find the solution.
(ii) We can express the initial value $\theta_0 \in [0, 1)$ as binary number

$$\theta_0 = \frac{b_0}{2} + \frac{b_1}{2^2} + \frac{b_2}{2^3} + \cdots = \sum_{j=0}^{\infty} \frac{b_j}{2^{j+1}}, \qquad b_j \in \{0, 1\}.$$

Let θ_0 given in binary as

$$\theta_0 = 0.b_0 b_1 \ldots b_7 = 0.10110101.$$

Find $\theta_1, \ldots, \theta_8$.

Solution 71. (i) We have $\theta_{t+1} = 2^t\theta_0 \mod 1$.
(ii) We obtain

$$\theta_1 = 0.0110101, \quad \theta_2 = 0.110101, \quad \theta_3 = 0.10101, \quad \theta_4 = 0.0101,$$

$$\theta_5 = 0.101, \quad \theta_6 = 0.01, \quad \theta_7 = 0.1, \quad \theta_8 = 0$$

Problem 72. Consider the map $f_n : [-1,1] \to [-1,1]$

$$f_n(x) = \cos(nx \arccos(x)), \qquad n = 1, 2, \ldots.$$

Find the fixed points and study their stability. Note that

$$\frac{d}{dx}\text{arcos}(x) = -\frac{1}{\sqrt{1-x^2}}.$$

Solution 72. From the equation to find the fixed points

$$x^* = \cos(nx^* \arccos(x^*))$$

we obtain

$$2k\pi \pm \arccos(x^*) = nx^* \arccos(x^*)$$

where k is an integer. If $k = 0$ and n is an odd integer, the fixed points are

$$x^*_{1,2} = \pm 1, \qquad x^*_{3,4} = \pm\frac{1}{n}$$

and some other fixed points that must be determined numerically. If $k = 0$ and n is not an odd integer, the fixed points are

$$x^*_1 = 1, \qquad x^*_{2,3} = \pm\frac{1}{n}$$

and some other fixed points that must be determine numerically. If $k \neq 0$, then no analytic expression exists for x^*. We have

$$\frac{df_n(x)}{dx} = -\sin(nx \arccos(x))\left(n \arccos(x) - \frac{nx}{\sqrt{1-x^2}}\right).$$

If $|df_n/dx|_{x=x^*} < 1$ the fixed point is stable. If $|df_n/dx|_{x=x^*} > 1$ the fixed point is unstable. If $|df_n/dx|_{x=x^*} = 1$ we find a first bifurcation. Consider $k = 0$ with $x^*_{1,2} = 1$. Note that $\arccos(1) = 0$. Then $df_n(x = 0)/dx = 0$. Thus this fixed point is stable.

Problem 73. Find an approximation for *Feigenbaum's universal constant* α in his equation

$$g(x) = -\alpha g(g(-x/\alpha)) \qquad g(0) = 1$$

by approximating g by a fourth degree polynomial. By construction

$$g : [-1/2, 1/2] \to [-1/2, 1/2].$$

Solution 73. Consider a (necessarily even) fourth-degree polynomial approximation to the *Feigenbaum function*

$$g(x) = 1 + ax^2 + bx^4. \qquad (1)$$

It follows that

$$g\left(-\frac{x}{\alpha}\right) = 1 + \frac{a}{\alpha^2}x^2 + \frac{b}{\alpha^4}x^4$$

and

$$\begin{aligned} g\left(g\left(-\frac{x}{\alpha}\right)\right) &= 1 + a\left(1 + \frac{a}{\alpha^2}x^2 + \frac{b}{\alpha^4}x^4\right)^2 + b\left(1 + \frac{a}{\alpha^2}x^2 + \frac{b}{\alpha^4}x^4\right)^4 \\ &= 1 + a\left[1 + \frac{2a}{\alpha^2}x^2 + \left(\frac{a^2+2b}{\alpha_4}\right)x^4 + \frac{2ab}{\alpha^6}x^6 + O(x^8)\right] \\ &\quad + b\left[1 + \frac{2a}{\alpha^2}x^2 + \left(\frac{a^2+2b}{\alpha_4}\right)x^4 + \frac{2ab}{\alpha^6}x^6 + O(x^8)\right]^2 \end{aligned}$$

Hence

$$g\left(g\left(-\frac{x}{\alpha}\right)\right) = 1 + a\left[1 + \frac{2a}{\alpha^2}x^2 + \left(\frac{a^2}{\alpha^4} + \frac{2b}{\alpha_4}\right)x^4 + \frac{2ab}{\alpha^6}x^6\right]$$

$$+b\left[1 + \frac{4a}{\alpha^2}x^2 + \left(\frac{4a^2}{\alpha^2} + 2\left(\frac{a^2+2b}{\alpha^4}\right)\right)x^4 + \left(\frac{4a}{\alpha^2}\left(\frac{a^2+2b}{\alpha^4}\right) + \frac{4ab}{\alpha^6}\right)x^6\right]$$

$$+O(x^8). \qquad (2)$$

Substitute (1) and (2) in the Feigenbaum equation yields

$$1 + ax^2 + bx^4 = -\alpha\left[(1+a+b) + \frac{2(a+2b)}{\alpha^2}x^2 + \frac{a^3 + 2ab + 6a^2b + 4b^2}{\alpha^4}x^4\right.$$

$$\left. + \frac{2b(a^2 + 4ab + 2a^3)}{\alpha^6}x^6 + O(x^8)\right]. \qquad (3)$$

Equate the coefficients of the lowest three powers of x on either side of the equation provides the system of three equations

$$a + b = -\frac{1+\alpha}{\alpha} \qquad (4)$$

$$2a + 4b = -\alpha \qquad (5)$$

$$a^3 + 2ab + 6a^2b + 4b^2 + \alpha^3 b = 0 \qquad (6)$$

for the unkowns a, b, α. Equations (4) and (5) can be solved to find a and b in terms of α

$$a = \frac{\alpha^2 - 4\alpha - 4}{2\alpha}, \quad b = \frac{2 + 2\alpha - \alpha^2}{2\alpha}. \qquad (7)$$

Substituting (7) in (6) and simplifying, a polynomial equation in α is obtained

$$4\alpha^7 - 3\alpha^6 - 60\alpha^5 + 104\alpha^4 + 168\alpha^3 - 240\alpha^2 - 384\alpha - 128 = 0.$$

Numerically, only three real roots are found, *viz.* $-3.9163\ldots, -0.7059\ldots$ and $2.5430\ldots$. Since $\alpha > 0$ by construction, then follows (to fourth decimal accuracy) that

$$\alpha = 2.5430. \tag{8}$$

If this value is substituted in (7), one finds that

$$a = -1.5222, \qquad b = 0.1276,$$

and (1) assumes the form $g(x) = 1 - 1.5222x^2 + 0.1276x^4$.

Problem 74. Consider the logistic equation

$$x_{t+1} = rx_t(1 - x_t), \qquad x_0 \in [0,1], \qquad t = 0, 1, 2, \ldots \tag{1}$$

Using the *renormalization technique* show that how the accumulation point (r_∞) and the structural universalities (δ and α) can be determined approximatively for (1).

Solution 74. The second iterate of (1) is

$$x_{t+2} = r^2 x_t - (r^2 + r^3)x_t^2 + 2r^3 x_t^3 - r^3 x_t^4. \tag{2}$$

For the fixed points x^* of (2), we have four solutions

$$x_0^* = 0, \qquad x_1^* = 1 - \frac{1}{r}, \qquad x_{2,3}^* = \frac{1}{2r}(r + 1 \pm \sqrt{(r-3)(r+1)}). \tag{3}$$

Now we introduce a convected transformation by setting

$$x_t = y_t + x_i \qquad \text{and} \qquad x_{t+2} = y_{t+1} + x_i \tag{4}$$

where x_i is either one of the solutions x_2 and x_3. Inserting this ansatz into (2) one finds after some algebra and neglecting higher order terms, that

$$\bar{x}_{t+1} = \bar{r}\bar{x}_t(1 - \bar{x}_t) \tag{5}$$

where

$$\bar{r} := 4 + 2r - r^2, \qquad \bar{x}_t = Cy_t, \qquad \bar{x}_{t+1} = Cy_{t+1} \tag{6}$$

and

$$C := \frac{r(r^2 - 2r - 3) \pm 3r\sqrt{(r-3)(r+1)}}{4 + 2r - r^2}. \tag{7}$$

The map (5) has the same form as the original one, and is thus self similar. Consequently, the same procedure can be repeated ad infinitum while only changing r according to $r \to 4 + 2r - r^2$. Thus we have

$$r_{n+1} = 4 + 2r_n - r_n^2. \tag{8}$$

Now as $n \to \infty$ we find $r_{n+1} = r_n = r_\infty$ which is the accumulation point where the chaotic region starts. This is calculated from

$$r_\infty = 4 + 2r_\infty - r_\infty^2$$

and found to be

$$r_{\infty_{1,2}} = \frac{1}{2}(1 \pm \sqrt{17}) = (-1.5615528, 2.5615528).$$

Evaluating the multiplier of internal iteration

$$r' := 2 - 2r_n = \delta_\infty$$

for $r_\infty = -1.5615528$, one finds that the *Feigenbaum constant* $\delta_\infty = r'$ is

$$\delta_\infty = r'|_{r_\infty} = 2 - 2(-1.5615528) \cong 5.123$$

Similarly, the rescaling constant is found to be $C = \alpha = -2.2399$. This is a rather good approximation to the exact values

$$r_\infty = -1.569946..., \qquad \delta = 4.669..., \qquad \alpha = -2.505....$$

Problem 75. Let $x \in [0,1]$. Then $4x(1-x) \in [0,1]$. Apply the transformation $x \mapsto 4x(1-x)$ to the differential one form

$$\alpha = \frac{1}{\pi}\frac{1}{\sqrt{x(1-x)}}dx.$$

Discuss.

Solution 75. First we have

$$dx \mapsto d(4x(1-x)) = 4d(x - x^2) = 4(1-2x)dx.$$

Now

$$x(1-x) \mapsto 4x(1-x)(1-4x(1-x)) = 4x(1-x)(1-2x)^2.$$

It follows that

$$\frac{dx}{\sqrt{x(1-x)}} \mapsto \frac{4(1-2x)dx}{2(1-2x)\sqrt{x(1-x)}} = \frac{2dx}{\sqrt{x(1-x)}}.$$

Thus $\alpha \mapsto 2\alpha$ (scaling) under the transformation. Note that $\ln(2)$ is the Liapunov exponent of the logistic map.

Problem 76. Consider a one-dimensional chaotic map

$$x_{t+1} = f(x_t), \qquad t = 0, 1, 2, \ldots$$

Then the *invariant measure* $\rho(x)$ with the *multiplication factor* m satisfies

$$\rho(x_{t+1})|dx_{t+1}| = m \cdot \rho(x_t)|dx_t|.$$

Thus

$$\left|\frac{df(x)}{dx}\right| = m \cdot \frac{\rho(x)}{\rho(f(x))}.$$

(i) Find the multiplication factor for the logistic map $f : [0,1] \to [0,1]$, $f(x) = 4x(1-x)$ with the invariant measure

$$\rho(x) = \frac{1}{\sqrt{x(1-x)}}.$$

(ii) Find the multiplication factor m for the *Baker map*

$$f_3(x) = \begin{cases} 3x & 0 \le x \le 1/3 \\ 2 - 3x & 1/3 \le x \le 2/3 \\ 3x - 2 & 2/3 \le x \le 1 \end{cases}$$

with the invariant measure $\rho(x) = 1$. The map f_3 is piecewise linear and continuous.

Solution 76. (i) We have

$$\rho(f(x)) = \frac{1}{\pi} \cdot \frac{1}{\sqrt{4x(1-x)(1-4x(1-x))}} = \frac{1}{2\pi} \cdot \frac{1}{\sqrt{x(1-x)(1-2x)^2}}$$
$$= \frac{1}{2\pi} \cdot \frac{1}{1-2x} \cdot \frac{1}{\sqrt{x(1-x)}}.$$

Thus

$$\frac{\rho(x)}{\rho(f(x))} = 2(1-2x).$$

Since $df(x)/dx = 4(1-2x)$ it follows that $m = 2$. Note that $\ln(m) = \ln(2)$ is the Liapunov exponent of the logistic map.
(ii) Since $\rho(f(x)) = \rho(x) = 1$ and $|df(x)/dx| = 3$ we obtain $m = 3$ with $\ln(m) = \ln(3)$ the Liapunov exponent.

Problem 77. A one-parameter family of analytic functions $f_r : \mathbb{R} \to \mathbb{R}$ is defined by

$$f_r(x) = -x(x + r), \qquad r > 0$$

where r is the bifurcation parameter.
(i) Find the fixed points of f_r.
(ii) Show that f_r has an infinite number of eventually fixed points in the interval $[-(1+r), 1]$ for $r > 2$.
Hint. It is sufficient to show that any $x \in [-(1+r), 1]$ has two preimages, both of which lie in $[-(1+r), 1]$.

Solution 77. (i) The fixed point equation $f_r(x^*) = x^*$ yields the two fixed points $x^* = 0$ and $x^* = -(1+r)$.
(ii) Let $x_0 \in [-(1+r), 1]$ and let x be its preimage. Then from $-x(x+r) = x_0$ we obtain

$$x_{1,2} = \frac{1}{2}\left(-r \pm \sqrt{r^2 - 4x_0}\right).$$

Next we show that x_1 and x_2 lie in the interval $[-(1+r), 1]$. We have

$$\begin{aligned} -(r+1) \leq x_0 \leq 1 \\ -4 \leq -4x_0 \leq 4(r+1) \\ r^2 - 4 \leq r^2 - 4x_0 \leq r^2 + 4r + 4 \\ 0 < r^2 - 4 \leq r^2 - 4x_0 \leq (r+2)^2. \end{aligned}$$

Hence

$$0 < \sqrt{r^2 - 4x_0} \leq r + 2.$$

It follows that

$$-\frac{r}{2} < \frac{1}{2}\left(-r + \sqrt{r^2 - 4x_0}\right) \leq 1$$

and therefore $x_1 \in [-r/2, 1] \subset [-(1+r), 1]$. Furthermore

$$\begin{aligned} -(r+2) \leq -\sqrt{r^2 - 4x_0} < 0 \\ -2(r+1) \leq -r - \sqrt{r^2 - 4x_0} < -r \\ -(r+1) \leq \frac{1}{2}(-r - \sqrt{r^2 - 4x_0}) < -\frac{r}{2}. \end{aligned}$$

Hence $x_2 \in [-(r+1), -r/2] \subset [-(r+1), 1]$. Therefore any point in $[-(r+1), 1]$ has two preimages in $[-(1+r), 1]$. In particular, each of the fixed points 0 and $-(r+1)$ has 2 preimages of order 1; 4 preimages of order 2; etc. Each of these two fixed points therefore has an infinite number of preimages; these are all eventually fixed.

Problem 78. Let $a, b, c, d \in \mathbb{R}$ and $ad - bc \neq 0$, $c \neq 0$. Consider the map $f : \mathbb{R} \to \mathbb{R}$

$$f(x) = \frac{ax + b}{cx + d}$$

or written as a difference equation

$$x_{t+1} = \frac{ax_t + b}{cx_t + d}, \quad t = 0, 1, \ldots$$

(i) Find a $v(x)$ such that

$$v(f(x)) = v(x)\frac{df(x)}{dx}.$$

(ii) Find the fixed points of f.

Solution 78. (i) Since

$$\frac{df(x)}{dx} = \frac{ad - bc}{(cx + d)^2}$$

we have to solve

$$v\left(\frac{ax + b}{cx + d}\right) = v(x)\frac{(ad - bc)}{(cx + d)^2}.$$

The solution is

$$v(x) = cx^2 + (d - a)x - b.$$

(ii) We have to solve quadratic equation

$$x = \frac{ax + b}{cx + d} \quad \Rightarrow \quad x^2 + \frac{(d - a)}{c}x - \frac{b}{c} = 0$$

to find the fixed points.

Problem 79. Let $x_0 = 1$. Study the nonlinear difference equation

$$x_{t+1} = 1 + 1/x_t, \quad t = 0, 1, 2, \ldots$$

First find the fixed points.

Solution 79. The two fixed points are

$$x_1^* = \frac{1}{2}(1 + \sqrt{5}), \quad x_2^* = \frac{1}{2}(1 - \sqrt{5}).$$

We have $x_0 = 1$, $x_1 = 2$, $x_2 = 3/2$, $x_3 = 5/3$. The sequence tends to the fixed points x_1^*. With the transformation $x_t = y_{t+1}/y_t$ we obtain

$$y_{t+2} = y_{t+1} + y_t, \quad t = 0, 1, 2, \ldots$$

with $y_0 = 1$, $y_1 = 1$, $y_2 = 2$, $y_3 = 3$, $y_4 = 5$ This is the difference equation for the Fibonacci sequence. The solution of this linear difference equation is

$$y_t = a_1 r_1^t + a_2 r_2^t$$

with $r_1 = x_1^*$, $r_2 = x_2^*$ and a_1, a_2 are given by $y_0 = 1 = a_1 + a_2$ and $y_1 = 1 = a_1 r_1 + a_2 r_2$.

Problem 80. The *thermodynamic formalism* is as follows. Consider a chaotic map f. We define the *partition function*

$$Z_n(f,\beta) := \sum_{Fix(z \in f^{(n)})} \exp\left[-\beta \ln |(f^{(n)})'(z)|\right] \tag{1}$$

where Fix denotes the number of fixed points of the iterated map $f^{(n)}$. Next we introduce the *free energy* $F(f,\beta)$ per unit time (or unit "site")

$$F(f,\beta) := -\beta^{-1} \lim_{n\to\infty} \frac{1}{n} \ell n(Z_n(f,\beta)). \tag{2}$$

We have introduced the (inverse) temperature β, so we can formally develop a thermodynamics of chaos following the equilibrium thermodynamics. The internal energy U is

$$U(\beta) \equiv \frac{(\partial \beta F(f,\beta))}{\partial \beta} = \langle \ln |f'| \rangle_\beta. \tag{3}$$

The partition function $Z_n(f,\beta)$ can be characterized by the variational principle

$$-\ln(Z_n(f,\beta)) = \min(\beta(LCN) - h(\mu)) \tag{4}$$

where LCN is the Liapunov characteristic number. $h(\mu)$ is the Kolmogorov-Sinai entropy of the invariant measure μ. Therefore U is the LCN for the map f. The entropy S is defined as

$$S(f,\beta) := \beta^2 \frac{\partial F(f,\beta)}{\partial \beta} = \beta(U(f,\beta) - F(f,\beta)). \tag{5}$$

If f exhibits chaotic behaviour supported by an absolutely continuous measure, $F(f,1) = 0$. Hence

$$S(1) = \langle \ln |f'| \rangle_1. \tag{6}$$

The Kolmogorov entropy in this case is equal to the LCN. Thus we may identify $S(1)$ with the Kolmogorov entropy, if g allows an absolutely continuous invariant measure. In the $\beta \to 0$ limit, we have

$$S(0) = \lim_{n\to\infty} \frac{1}{n} \ln(\#\mathrm{Fix}(f^{(n)}))$$

which is the formula for the *topological entropy* for g. *Dinaburg's theorem* asserting the inequality, topological entropy $\geq$ Kolmogorov entropy, has the following expression

$$S(0) \geq S(1).$$

This is obvious since entropy increases as β decreases due to the positivity of the "heat capacity". We identify $S(\beta_c)$ with the Kolmogorov entropy, where β_c is the β such that $F(f, \beta_c) = 0$.
(i) Let $f_p : [0,1] \to [0,1]$ be a map defined by

$$f_p(x) = \begin{cases} x/p & x \in [0,p] \\ (1-x)/(1-p) & x \in [p,1]. \end{cases}$$

Find $Z_n(f_p, \beta)$. The fixed points of f_p are $x^* = 0$ and $x^* = 1/(2-p)$.
(ii) Find $S(1)$.

Solution 80. (i) We obtain for the partition function

$$Z_n(f_p, \beta) = (p^\beta + q^\beta)^n \qquad q := 1 - p$$

so that

$$\begin{aligned} F(f_p, \beta) &= -\beta^{-1} \ln(p^\beta + q^\beta) \\ U(f_p, \beta) &= -(p^\beta \ln(p) + q^\beta \ln(q))/(p^\beta + q^\beta) \\ S(f_p, \beta) &= \ln(p^\beta + q^\beta) - \beta(p^\beta \ln(p) + q^\beta \ln(q))/(p^\beta + q^\beta). \end{aligned}$$

(ii) We find

$$S(1) = -p\ln(p) - q\ln(q), \qquad S(0) = \ln(2).$$

If p is irrational, then there is no Markov partition. The limit $\beta \to \infty$ should correspond to the absolute zero of temperature. Then the contribution to the partition function comes only from the most stable orbits ("ground states"). The Gibbs distribution is concentrated on 0 if $p > q$ and on the other fixed point of f_p if $p < q$. Since the distribution is atomic, the entropy $S(\infty)$ should be zero in these cases. Show that if $p = q = 1/2$, all the orbits have the same stability, so the "ground state" is degenerated. We have $S(\infty) = \ln(2)$.

Problem 81. Consider the *Chebyshev polynomials*

$$f(x) = 2x^2 - 1, \qquad g(x) = 4x^3 - 3x.$$

Find the "commutator"

$$(f \circ g)(x) - (g \circ f)(x) \equiv f(g(x)) - g(f(x)).$$

Apply computer algebra.

Solution 81. A SymbolicC++ program that will do the job is

```
// Chebyshev.cpp

#include <iostream>
#include "symbolicc++.h"
using namespace std;

Symbolic f(const Symbolic& x) { return 2*x*x - 1; }
Symbolic g(const Symbolic& x) { return 4*x*x*x - 3*x; }

Symbolic C(Symbolic (*f)(const Symbolic&),Symbolic (*g)(const Symbolic&),
           const Symbolic &x)
{ return (f(g(x))-g(f(x))); }

int main(void)
{
  Symbolic x("x");
  Symbolic result = C(f,g,x);
  cout << "result = " << result << endl;
  double v = 2.5;
  double resultvalue = result[x==v];
  cout << "resultvalue = " << resultvalue << endl;
  return 0;
}

// The output is
// result = 0
// resultvalue = 0
```

Problem 82. Consider the one-dimensional map (logistic map) $f : [0,1] \to [0,1]$

$$f(x) = 4x(1-x).$$

A computational analysis using a finite state machine with base 2 arithmetic in fixed point operation provides one-dimensional maps with a lattice of 2^N sites labelled by numbers

$$x = \sum_{j=1}^{N} \frac{\epsilon_j}{2^j}, \qquad \epsilon_j \in \{0, 1\}$$

and N defines the machine's precision. Consider $N = 8$ bits and $x = 1/8$. Calculate the orbit $f(x)$, $f(f(x))$, $f(f(f(x)))$, ... with this precision. Discuss. Apply the `Rational` class of SymbolicC++ and `algorithm`, `list` from the Standard Template Library.

The problem can also be formulated as follows. Consider the logistic map $f(x) = 4x(1-x)$. If $x \in \{0, 2^{-n}, \ldots, 1-2^{-n}\}$ is stored in an n-bit register representing numbers $\{0, 2^{-n}, \ldots, 1-2^{-n}\}$ then the computation of f introduces two forms of truncation: if $f(x) \geq 1$ then $f(x)$ is truncated to $1-2^{-n}$ and if $f(x) < 1$ then $f(x)$ must be truncated to the closest $y \in \{0, 2^{-n}, \ldots, 1-2^{-n}\}$ with $y \leq f(x)$. Of course, instead of truncation we could also round to the nearest value that can be stored in the register or use another strategy. The truncation can be expressed as follows:

$$g(x) = \begin{cases} 1-2^{-n} & x \geq 1 \\ 2^{-n}\lfloor 2^n x \rfloor & \text{otherwise} \end{cases}.$$

Since the n-bit register can only store 2^n different numbers, the sequence $(g \circ f)(x), (g \circ f)^2(x), \ldots$ must be eventually periodic. Write a C++ program which find the periodic subsequence of the sequence generated by x for $n = 8$.

Solution 82. The bit representation of 1/8 with $N = 8$ is `00100000`. Now $f(1/8) = 7/16$ with the bit representation `01110000` with no rounding errors. Next we have $f(f(1/8)) = 63/64$ with the bit representation `11111100`. The following C++ program finds the eventually periodic orbit.

```
/* periodic.cpp */

#include <algorithm>
#include <iostream>
#include <list>
#include "rational.h"

template <class T>
Rational<T> floor(const Rational<T> &r) { return r-r.frac(); }

template <class T>
T gof(const T& x) { return floor(256*4*x*(1-x))/256; }

template <class T>
void periodic_sequence(const T &x0)
{
 int p = 0, n = 0;
 T x = x0;
 list<T> s;
 typename list<T>::iterator si;
 // iterate until we find a periodic subsequence
 while((si=find(s.begin(),s.end(),x))==s.end())
 {
 s.push_back(x); cout << x << endl;
 x = gof(x); ++n;
```

```
 }
 s.push_back(x);
 cout << " --- " << endl;
 while(si != s.end()) { cout << *(si++) << endl; ++p; }
 --p;
 cout << endl;
 cout << "Periodic sequence, length " << p << " after "
      << n-p << " iterations." << endl;
 cout << endl;
}

int main(void)
{
 periodic_sequence(Rational<int>("1/8"));
 periodic_sequence(1/8.0);
 return 0;
}
```

The output is

```
1/8
7/16
63/64
15/256
7/32
175/256
221/256
15/32
255/256
3/256
11/256
21/128
35/64
253/256
 ---
11/256
21/128
35/64
253/256
11/256
Periodic sequence, length 4 after 10 iterations.

0.125
0.4375
0.984375
0.0585938
```

```
0.21875
0.683594
0.863281
0.46875
0.996094
0.0117188
0.0429688
0.164062
0.546875
0.988281
---
0.0429688
0.164062
0.546875
0.988281
0.0429688
Periodic sequence, length 4 after 10 iterations.
```

1.2.2 Supplementary Problems

Problem 1. Consider the nonlinear difference equation

$$x_{t+1} = \frac{x_t}{1+x_t}, \qquad t = 0, 1, 2, \ldots \tag{1}$$

with $x_0 > 0$ as a prescribed positive number (initial value).
(i) Show that there is only one fixed points $x^* = 0$ of the difference equation. Show that the fixed point is stable.
(ii) Show that if $x_0 \neq 0$, then $x_t \neq 0$ for every t-value.
(iii) Let

$$v_t = \frac{1}{x_t}, \qquad t = 0, 1, 2, \ldots. \tag{2}$$

Show that (1) is transformed into the linear difference equation

$$v_{t+1} = v_t + 1, \qquad t = 0, 1, 2, \ldots. \tag{3}$$

(iv) Show that $v_t = v_0 + t$ is the solution of the initial value problem of (3).
(v) Then use (2) to find the solution of the initial value problem of the original difference equation. Show that the solution is given by

$$x_t = \frac{x_0}{1+tx_0}, \qquad t = 0, 1, 2, \ldots.$$

Problem 2. Let $a > 0$ and $n = 2, 3, \ldots$. Obtain the general solution of the equation

$$x_{t+1} = a x_t^n, \quad t = 0, 1, 2, \ldots$$

in the form

$$x_t = c^{n^t} a^{1/(1-n)}$$

where c is an arbitrary constant. Take the logarithm of both sides. Show that this result can also be written in the form

$$x_t = a_0^{n^t} a^{(1-n^t)/(1-n)}.$$

Problem 3. Study the nonlinear difference equation

$$e^{4x_{t+1}} = \frac{e^{8x_t} - e^{4x_t} + 4}{e^{4x_t} + 3}, \quad t = 0, 1, 2, \ldots$$

and $x_0 \geq 0$. The only fixed points are 0 and ∞. Let $x_0 = 1$. Find x_1, x_2, $\ldots$.

Problem 4. The logistic map $x_{t+1} = 4x_t(1 - x_t)$ is well studied. Study the map

$$\sinh(x_{t+1}) = 4x_t(1 - x_t), \quad t = 0, 1, \ldots.$$

Is $x^* = 0$ a fixed point? Note that

$$\sinh(x) := \sum_{j=0}^{\infty} \frac{x^{2j+1}}{(2j+1)!}$$

and $f : \mathbb{R} \to \mathbb{R}$, $f(x) = \sinh(x)$ is a diffeomorphism.

Problem 5. Study the one-dimensional map

$$x_{t+1} = \frac{1}{2} \ln(\cosh(4x_t))$$

with $t = 0, 1, \ldots$ and the three initial values $x_0 = 1/2, 1, 2$.

Problem 6. Consider the logistic map $f : [0, 1] \to [0, 1]$, $f(x) = 4x(1-x)$. Calculate

$$g(n) = \int_0^1 x^{n-1} f(x) dx, \quad n = 1, 2, \ldots$$

Reconstruct the function f from $g(n)$.

Problem 7. Prove the following theorem: If a continuous function of the real numbers has a periodic point with prime period three, then it has periodic points of all prime periods.

Problem 8. Consider the Hilbert space of square integrable functions $L_2([-1,1])$. Then the chaotic map $f : [-1,1] \to [-1,1]$

$$f(x) = 1 - 2x^2$$

is an element of $L_2([-1,1])$. Using the *Legendre polynomials* we can form an orthonormal basis

$$B = \left\{ \phi_\ell(x) = \frac{\sqrt{2\ell+1}}{\sqrt{2}} \frac{1}{2^\ell \ell!} \frac{d^\ell}{dx^\ell}(x^2-1)^\ell \, : \, \ell = 0,1,2,\ldots \right\}$$

in $L_2([-1,1])$. Note that $\bar{\phi}_\ell(x) = \phi_\ell(x)$. Consider the infinite dimensional matrix $F = (F_{jk})$ $(j,k = 0,1,2,\ldots)$

$$F_{jk} = \langle \phi_j | f(x) | \phi_k \rangle = \int_{-1}^{+1} \phi_j(x) f(x) \phi_k(x), \quad j,k = 0,1,2,\ldots$$

which acts as a linear bounded operator in the Hilbert space $\ell_2(\mathbb{N}_0)$. How can one reconstruct the function f from the natrix F and the orthonormal basis? Calculate the matrix F and find the spectrum of F.

Problem 9. Show that the only real solution of $\sin(x) = x$ is $x = 0$, i.e. $f(x) = \sin(x)$ only admits the fixed point $x^* = 0$.

Problem 10. Consider the analytic function $f : \mathbb{R} \to \mathbb{R}$

$$f(x) = e^{-x} + 1.$$

(i) Show that $f(x^*) = x^*$ (fixed point equation) has only one real solution.
(ii) Show that this fixed point is given by

$$x^* = 1 + \sum_{j=1}^{\infty} \frac{(-1)^{j-1} j^{j-1} e^{-j}}{j!}.$$

Problem 11. Consider the polynomial $T_5 : \mathbb{R} \to \mathbb{R}$

$$T_5(x) = 16x^5 - 20x^3 + 5x$$

which is one of the *Chebyshev polynomials* of first kind.
(i) Show that if $x \in [-1,1]$, then $T_5(x) \in [-1,1]$.
(ii) Find the fixed points in $[-1,1]$ and show that they are unstable. First derive the variational equation.
(iii) Find the critical points of T_5 in $[-1,1]$ and study symbolic dynamics.

(iv) Find the exact solution of

$$x_{t+1} = 16x_t^5 - 20x_t^3 + 5x_t.$$

(v) Find the Liapunov exponent.

Problem 12. Show that the analytic map $f : [-1, 1] \to [-1, 1]$

$$f(x) = 4x^3 - 3x$$

and the continuous map $g : [-1, 1] \to [-1, 1]$

$$g(x) = \begin{cases} 4x + 3 \text{ for } -1 \le x \le -1/2 \\ -2x \text{ for } -1/2 \le x \le 1/2 \\ 4x - 3 \text{ for } 1/2 \le x \le 1 \end{cases}$$

are topologically conjugate.

Problem 13. Let $r \ge 0$ be the bifurcation parameter and $x_0 \ge 0$. Study the map

$$x_{t+1} = x_t^5 e^{-x_t} + r, \quad t = 0, 1, 2, \ldots$$

Since $x_0 \ge 0$ we have $x_t \ge 0$ for all t. First find the fixed points and study their stability.

Problem 14. Consider the function $f : \mathbb{R} \to \mathbb{R}$, $f(x) = 2x^3$. Then the inverse function f^{-1} is given by

$$f^{-1}(x) = \sqrt[3]{\frac{x}{2}}.$$

Find the fixed points of f and f^{-1}. Discuss.

Problem 15. Consider the analytic function $f : \mathbb{R} \to \mathbb{R}$

$$f(x) = x + \sin(x).$$

(i) Show that f admits infinite many fixed points given by $x^* = n\pi$, where $n \in \mathbb{Z}$.
(ii) Study the stability of these fixed points.
(iii) Show that the function is a diffeomorphism. Find f^{-1}. Find the inverse of f applying the *Lagrange inversion theorem.* Let $y = f(x)$, where f is analytic at a point p and $df(x = p)/dx \ne 0$. Then one can invert on a neighbourhood of $f(p)$, i.e. $x = g(y)$, where the function g is analytic at the point $f(p)$. The series expansion is

$$g(y) = p + \sum_{j=1}^{\infty} \left(\lim_{x \to p} \left(\frac{(y - f(p))^j}{j!} \frac{d^{j-1}}{dx^{j-1}} \left(\frac{x - p}{f(x) - f(p)} \right)^j \right) \right).$$

In the present case we have $p = 0$. Then $df(x = p)/dx = 2$ and $f(p = 0) = 0$ and the series expansion simplifies to

$$g(y) = \sum_{j=1}^{\infty} \left(\lim_{x \to p} \left(\frac{y^j}{j!} \frac{d^{j-1}}{dx^{j-1}} \left(\frac{x}{f(x)} \right)^j \right) \right).$$

Problem 16. Show that the analytic function $f : \mathbb{R} \to \mathbb{R}$

$$f(x) = -2x - \sin(x)$$

is a diffeomorphism. Show that $x^* = 0$ is a fixed point. Is the fixed point stable?

Problem 17. (i) Show that the logistic map $x_{t+1} = 4x_t(1 - x_t)$ can be written in matrix form as

$$x_{t+1} = \begin{pmatrix} x_t & 1 - x_t \end{pmatrix} \begin{pmatrix} 0 & 2 \\ 2 & 0 \end{pmatrix} \begin{pmatrix} x_t \\ 1 - x_t \end{pmatrix}.$$

(ii) Show that the logistic map $x_{t+1} = 4x_t(1 - x_t)$ can be written as

$$\begin{pmatrix} x_{t+1} \\ 1 - x_{t+1} \end{pmatrix} = \begin{pmatrix} 0 & 2 & 2 & 0 \\ 1 & -1 & -1 & 1 \end{pmatrix} \left(\begin{pmatrix} x_t \\ 1 - x_t \end{pmatrix} \otimes \begin{pmatrix} x_t \\ 1 - x_t \end{pmatrix} \right)$$

where $\otimes$ denotes the Kronecker product.

Problem 18. (i) Construct a polynomial $p(x) = x^2 + ax + b$ that admits the roots

$$\frac{\sqrt{3} + 3}{6}, \quad -\frac{\sqrt{3} - 3}{6}.$$

(ii) Construct a polynomial $p(x) = x^2 + ax + b$ that admits the fixed points

$$\frac{\sqrt{30} + 6}{6}, \quad -\frac{\sqrt{30} - 6}{6}.$$

Problem 19. Let $x > 0$. Find the fixed points of the functions $f_1(x) = x + 1/x$, $f_2(x) = x - 1/x$ and study their stability.

Problem 20. Study the map $f : [0, 1) \to [0, 1)$

$$f(x) = \begin{cases} 2x \pmod 1 & 0 \le x < 1/2 \\ 4x \pmod 1 & 1/2 \le x < 1 \end{cases}$$

(i) Find the fixed points.
(ii) Show that f preserves Lebesgue measure.
(iii) Does the map show chaotic behaviour?

Problem 21. (i) Consider the analytic function $f : \mathbb{R} \to \mathbb{R}$

$$f(x) = \frac{\pi}{2} + x - \arctan(x).$$

Show that the function has no fixed points. Show that

$$|f(x) - f(y)| < |x - y|$$

for all $x \neq y$.
(ii) Let $r > 0$. Show that the analytic function $g_r : \mathbb{R} \to \mathbb{R}$

$$g_r(x) = rx + \arctan(x)$$

is a diffeomorphism. Show that $dg_r/dx > 0$ for all $x \in \mathbb{R}$. Note that

$$\frac{d}{dx} \arctan(x) = \frac{1}{1+x^2}.$$

Problem 22. Let $r > 0$. Consider the *Newton map* $f_r : (1, \infty) \to \mathbb{R}$ given by

$$f_r(x) = \frac{1}{2}\left(x + \frac{r}{x}\right).$$

(i) Show that if $r \in (1,3)$, then f_r maps $(1, \infty)$ into itself, i.e. $f(x) \in (1, \infty)$ for all $x \in (1, \infty)$.
(ii) Show that the map f_r is a contraction if $r \in (1,3)$. Find the fixed points as a function of the parameter r.

Problem 23. Demonstrate the existence of an orbit of $f : [0,1] \to [0,1]$, $f(x) = 4x(1-x)$ with prime period three.

Problem 24. (i) Consider the *Bernoulli map* $f : [0,1) \to [0,1)$ given by

$$f(x) = 2x \bmod 1.$$

Find a point $x_0 \in [0,1)$ whose orbit is dense in $[0,1)$.
(ii) A generalization of the Bernoulli map is the map

$$f(x) = Dx \bmod 1$$

with $D \geq 2$. Show that the Liapunov exponent for almost all initial values is given by $\ln(D)$.

Problem 25. Let $r > 0$ and f_r be defined by

$$f_r(x) := \begin{cases} x/r & x \in [0,r] \\ 1 & x \in [r, 1-r] \\ (1-x)/r & x \in [1-r, r] \end{cases}$$

Show that this allows a *Cantor set* C as a maximal invariant set. Show that the Hausdorff dimension D_H of C is $-\ln(2)/\ln(r)$. Show that there is an invariant measure with the Kolmogorov entropy $\ln(2)$ supported on C. Show that the *partition function* is

$$Z_n(f_r, \beta) = (2r^\beta)^n.$$

Show that from $Z_n(f_r, \beta)$ we obtain

$$F(f_r, \beta) = -\ln(r) - \beta \ln(2)$$

which implies $F(f_r, D_H) = 0$. Show that $S(f_r, \beta)$ is $\ln(2)$. Show that $S(\infty) = \ln(2)$.

Problem 26. Let f be a continuous map from the unit interval $[0,1]$ onto itself, i.e. $f([0,1]) = [0,1]$.
(i) Show that the map f must have at least one fixed point.
(ii) Show that $f^{(2)}$ must have at least two fixed points.

Problem 27. Consider a continuous map $f : [0,1] \to [0,1]$. Show that there are points in $[0,1]$ that are not fixed points, periodic points, or eventually periodic points of the map f.

Problem 28. Let a sequence of functions $f_k(x)$ be defined as follows. The zeroth function if defined to be $f_0(x) = (x)$, the first function to be $f_1(x) = a_1/(b_1 + x)$ and the kth function is obtained from the preceding function $f_{k-1}(x)$ by replacing x by $a_k/(b_k + x)$, where $a_1, a_2, \cdots$ and $b_1, b_2, \cdots$ are constants.
(i) Show that

$$f_0(x) = x, \qquad f_1(x) = \frac{a_1}{b_1 + x}, \qquad f_2(x) = \frac{a_1}{b_1 + \frac{a_2}{b_2 + x}}, \cdots \tag{1}$$

and, in general,

$$f_k(x) = \cfrac{a_1}{b_1 + \cfrac{a_2}{b_2 + \cfrac{a_3}{b_3 + \cfrac{\ddots}{\ddots + \cfrac{a_k}{b_k + x}}}}}.$$

The expression $f_k \equiv f_k(0)$, obtained by setting $x = 0$ in $f_k(x)$, is called a continued fraction of k stages.
(ii) Show that the result of clearing fractions in the expression for $f_k(x)$ is the ratio of two linear functions of x, of the form

$$f_k(x) = \frac{A_k + C_k x}{B_k + D_k x}, \qquad k = 0, 1, 2, \cdots,$$

Problem 29. Consider the symmetric tent map ($t = 0, 1, \ldots$)

$$x_{t+1} = \begin{cases} 2x_t & x_t \in [0, 1/2] \\ 2(1 - x_t) & x_t \in [1/2, 1] \end{cases}$$

Derive the master equation

$$\rho_n(x) = \frac{1}{2}\left(\rho_{n-1}\left(\frac{x}{2}\right) + \rho_{n-1}\left(1 - \frac{x}{2}\right)\right), \qquad n = 1, 2, \ldots$$

given some initial non-equilibrium density $\rho_0(x)$.

Problem 30. Let $\alpha = \sqrt{2}$. Consider the map

$$x_j = (j+1)^2 \alpha \mod 1$$

where $j = 0, 1, \ldots$. The sequence x_0, x_1, x_2, $\ldots$, is uniformly distributed on the unit interval (equidistribution theorem). Study numerically the Liapunov exponent and Hurst exponent for this map. Study also the case where $\alpha = (\sqrt{5} - 1)/2$.

Problem 31. Let $n = 0, 1, 2, \ldots$. The *Fermat numbers* are given by

$$F_n = 2^{(2^n)} + 1.$$

Show that the Fermat numbers satisfy the recurrence relation

$$F_{n+1} = (F_n - 1)^2 + 1$$

with $F_0 = 3$.

Problem 32. Consider the analytic function $f : \mathbb{R} \to \mathbb{R}$

$$f(x) = 1 + x + \cos(x)\cosh(x).$$

Show that the fixed point equation $f(x^*) = x^*$ has infinitely many solutions.

Problem 33. Consider the function $f : \mathbb{R} \to \mathbb{R}$

$$f(x) = |1 - |2 - |3 - x|||.$$

(i) Find the fixed points. Are the fixed points stable? Obviously 0 and 1 are fixed points.
(ii) Is the function continuous?
(iii) Find minima and maxima of f.
(iv) Find $f(5)$, $f(f(5))$, $f(f(f(5)))$. Discuss.

Problem 34. Study the difference equation

$$x_{t+1} = e^{i\pi t}(1 - 2x_t^2), \quad t = 0, 1, 2, \ldots$$

where $x_0 \in [-1, 1]$. Note that $\cos(\pi) = -1$.

Problem 35. Consider the analytic function $f : [0, 1] \to [0, 1]$

$$f(x) = \frac{1}{2} - \frac{1}{2}\sin(2\pi x).$$

Find the fixed points and study their stability. The fixed $x^* = 1/2$ is obvious. The others two must be found numerically.

Problem 36. Consider the function $f : [0, 1] \to [0, 1]$

$$f(x) = 1 - \frac{1}{2}x^2 - \frac{1}{2}x^4.$$

Note that $f(0) = 1$ and $f(1) = 0$. If $x_1 < x_2$, then $f(x_1) > f(x_2)$. The function admits one fixed point. Find the fixed point and study its stability.

Problem 37. (i) Consider (*Catalan numbers*)

$$C_n = \frac{4n - 2}{n + 1} C_{n-1}, \quad n \geq 1$$

with $C_0 = 1$. Find C_1, C_2, C_3, C_4, C_5. Write a SymbolicC++ program utilizing the class `Verylong` to find these numbers. Write a Java program utilizing the class `BigInteger` to find these numbers.
(ii) Consider

$$C_n = \sum_{j=0}^{n-1} C_j C_{n-j-1}, \quad n \geq 1$$

with $C_0 = 1$. Find C_1, C_2, C_3, C_4, C_5. Write a SymbolicC++ program utilizing the class `Verylong` to find these numbers. Write a Java program utilizing the class `BigInteger` to find these numbers.

(iii) Let $n \geq 3$. Consider

$$T_n = \sum_{j=2}^{n-1} T_j T_{n-j+1}, \quad n \geq 3$$

with $T_2 = 1$. Find T_3, T_4, T_5, T_6. Write a SymbolicC++ program utilizing the class `Verylong` to find these numbers. Write a Java program utilizing the class `BigInteger` to find these numbers.

Problem 38. Consider one-dimensional smooth maps f and g. Show that the Liapunov exponents are invariant under conjugation. Let

$$h : x \to y, \quad x_{t+1} = g(x_t), \quad y_{t+1} = f(y_t)$$

and (*Liapunov exponents*)

$$\lambda(g) = \lim_{T\to\infty} \frac{1}{T} \sum_{t=0}^{T-1} \ln |g'(x_t)|, \qquad \lambda(f) = \lim_{T\to\infty} \frac{1}{T} \sum_{t=0}^{T-1} \ln |f'(y_t)|.$$

Problem 39. The *discrete Fourier transform* is used when a set of sample function values, $x(i)$, are available at equally spaced time intervals numbered $i = 0, 1, \ldots, N-1$. The discrete Fourier transform maps the given function values into the sum of a discrete number of sine and cosine waves whose frequencies are numbered $k = 0, 1, \ldots, N-1$, and whose amplitudes are given by

$$\hat{x}(k) = \frac{1}{N} \sum_{l=0}^{N-1} x(l) \exp\left(-i2\pi k \frac{l}{N}\right). \tag{1}$$

The equation can be written as

$$\hat{x}(k) = \frac{1}{N} \sum_{l=0}^{N-1} x(l) \cos\left(2\pi k \frac{l}{N}\right) - \frac{i}{N} \sum_{l=0}^{N-1} x(l) \sin\left(2\pi k \frac{l}{N}\right). \tag{2}$$

The inverse transformation is given by

$$x(l) = \sum_{k=0}^{N-1} \hat{x}(k) \exp\left(i2\pi k \frac{l}{N}\right). \tag{3}$$

To find (3), we use the fact that

$$\sum_{k=0}^{N-1} \exp\left(i2\pi k \frac{n-m}{N}\right) = N\delta_{nm} \tag{4}$$

where n, m are integers. Consider the symmetric *tent map*

$$f(x) := \begin{cases} 2x & \text{for } x \in [0,1/2] \\ 2-2x & \text{for } x \in (1/2,1] \end{cases}$$

(i) Show that the initial value $x = 6/17$ leads to the periodic orbit.
(ii) Find the Fourier transform of this orbit.

Problem 40. Consider the map $f : [0,1] \to [0,1]$

$$f(x) = \begin{cases} 4x & \text{for } \ 0 \le x \le 1/4 \\ -x/4+2 & \text{for } \ 1/4 \le x \le 1/2 \\ 2x-1 & \text{for } \ 1/2 \le x \le 3/4 \\ -2x+2 & \text{for } \ 3/4 \le x \le 1 \end{cases}$$

Does the map show chaotic behaviour?

Problem 41. Let $f_r(x) = rx(1-x)$ with $r > 2+\sqrt{5}$. Show that the Liapunov exponent of any orbit that remains in $[0,1]$ is greater than 0 if it exists.

Problem 42. Consider the map $f : [0,\infty) \to [0,\infty)$. Assume that f is analytic and $f(0) = 0$. Let $p > 0$ be a fixed point such that $f'(p) \ge 0$. Furthermore assume that $f'(x)$ is decreasing. Show that all positive x_0 converges to the fixed point p under the iteration of the map f.

Problem 43. To construct the *symbolic dynamics* of a dynamical system, the determination of the partition and the ordering rules for the underlying symbolic sequences is necessary. For one-dimensional mappings, the partition is composed of all the critical points. Consider the antisymmetric cubic map $f : [-1,1] \to [-1,1]$ $(t = 0,1,2,\ldots)$

$$x_{t+1} = f(x_t) \equiv rx_t^3 + (1-r)x_t, \qquad r \in [1,4]. \tag{1}$$

(i) Find the critical points of f. We denote the critical points by C and $\bar{C}$.
(ii) Show that the ternary partition marked by $\bar{C}$ and C divides the interval $[-1,1]$ into three monotonic branches.
(iii) The right branch to C is assigned 0, the left branch to $\bar{C}$ is assigned 2, whereas the part between $\bar{C}$ and C is 1. Show that nearly all trajectories are unambiguously encoded by infinite strings of bits $S(x) = (s_1 s_2 \cdots)$, where s_i is either 0, 1 or 2.
(iv) Referring to the natural ordering of the real numbers in the interval $[-1,1]$, show that the ordering rules for these symbolic strings can be defined, that is, considering two symbolic strings $S(x_1)$ and $S(x_2)$ from the initial points x_1 and x_2, $S(x_1) \ge S(x_2)$ if and only if $x_1 > x_2$.

(v) Find the kneading sequence K_g and K_s (i.e. the forward symbolic sequences from the maximal and minimal values $\bar{C}$ and C).

Problem 44. Consider the one-dimensional map $f : [0,1] \to [0,1]$

$$f(x) = \begin{cases} 1/2 - 2x & \text{if } 0 \leq x \leq 1/4 \\ -1/2 + 2x & \text{if } 1/4 \leq x \leq 3/4 \\ 5/2 - 2x & \text{if } 3/4 \leq x \leq 1 \end{cases}$$

Show that the invariant measure is given by

$$\rho(x) = \begin{cases} \epsilon & \text{if } 0 \leq x \leq 1/2 \\ 2 - \epsilon & \text{if } 1/2 \leq x \leq 1 \end{cases}$$

is invariant, where $\epsilon \in (0,1)$.

Problem 45. Consider the skew-symmetric tent map

$$f_r(x) = \begin{cases} rx & \text{for } 0 \leq x \leq 0.5 \\ r(1-x) & \text{for } 0.5 < x \leq 1 \end{cases}$$

where $r = \frac{1}{2}(1+\sqrt{5})$ (golden mean number). For *symbolic dynamics* we partition the unit interval $[0,1]$ into $L = [0, 0.5)$ and $R = (0.5, 1]$. Show that this tent map generates a first order Markov string with stationary probabilities

$$p(L) = \frac{1}{1+r^2}, \qquad p(R) = \frac{r^2}{1+r^2}.$$

Problem 46. Consider the analytic function $f : \mathbb{R} \to \mathbb{R}$

$$f(x) = 1 - (1-x^2)^2.$$

Find the fixed points and study their stability. Show that a fixed point is given by

$$x^* = \frac{1}{2}(\sqrt{5} - 1).$$

Problem 47. Consider the map (so-called *Mixmaster return map*)

$$x_{n+1} = f(x_n) \equiv x_n^{-1} - \lfloor x_n^{-1} \rfloor \qquad (1)$$

where $f(0) \equiv 0$ and $f : [0,1] \to [0,1]$. In analytic form this return map is the single-valued function,

$$f(x) = x^{-1} - k, \quad (k+1)^{-1} < x < k^{-1}; \quad k \in \mathbb{Z}^+. \qquad (2)$$

The function possesses an infinite number of discontinuities and is not injective since each x_0 has a countable infinity of inverse images, one on each interval $[(k+1)^{-1}, k^{-1}]$ for integral k.
(i) Show that the return mapping is *expansive*,

$$|f'(x)| > 1$$

on $x \in (0,1)$, everywhere.
(ii) Show that all the fixed points ($f(x^*) = x^*$) are unstable.
(iii) Show that the invariant measure is given by

$$\mu(x) = \frac{1}{(1+x)\ln(2)}.$$

(iv) Show that the metric entropy is given by

$$h(\mu, f) = -\frac{2}{(\ln 2)^2}\int_0^1 \frac{\ln(x)dx}{1+x} = \frac{\pi^2}{6(\ln(2))^2}.$$

It can be shown that f is mixing and therefore isomorphic to a Bernoulli shift.

Problem 48. Consider the unimodal map $f : [0,1] \to [0,1]$. We assume that f is continuous and reaches its maximal value at an interior point c of I. The point c is called the critical point of f. In both subintervals divided by c, $[0, c)$ and $(c, 1]$, the map f is strictly monotonic. We assume that $f(0) = f(1) = 0$. A discrete dynamical system

$$x_{t+1} = f(x_t), \qquad t = 0, 1, 2, \ldots$$

is defined from f by iteration. Given a starting point $x_0 \in I$, we use the notation

$$(x_0, x_1, \cdots, x_t, \cdots)$$

to denote the orbit from x_0. Using the *coarse-grained description*

$$A(x) := \begin{cases} 0 & \text{for } x < c \\ c & \text{for } x = c \\ 1 & \text{for } x > c \end{cases}$$

we transform the orbit $(x_0, x_1, x_2, \ldots)$ into an itinerary, that is,

$$I(x) = (A(x_0), A(x_1), \ldots, A(x_t), \ldots)$$

which is an infinite string over the alphabet $\{0, c, 1\}$. The kneading sequence of the unimodal map f is the itinerary $I(f(c))$, which decides nearly all other itineraries a given map f can have.

(i) Show that the logistic map $f(x) = 4x(1-x)$ is a unimodular map.
(ii) Obviously, $c = 1/2$ is a critical point of the map. Find a preimage of the map.
(iii) Find the orbit in the coarse-grained description for the initial value $x_0 = 1/3$.
(iv) Find the kneading sequence.

Problem 49. (i) Consider the nonlinear difference equation

$$y_{t+1} = 2y_t^2 - 1 \tag{1}$$

where $t = 0, 1, 2, \ldots$. Show that the substitution $y_t = \cos(x_t)$ reduces (1) to the equation

$$\cos(x_{t+1}) = \cos(2x_t).$$

Thus deduce that either $x_{t+1} = 2x_t + 2m\pi$ or $x_{t+1} = -2x_t + 2n\pi$, where m and n are arbitrary integers.
(ii) Show that from the second alternative we obtain the solution

$$y_t = \cos\left(2^t\theta + (-1)^t \frac{2n\pi}{3}\right)$$

where θ is an arbitrary constant and n an arbitrary integer.

Problem 50. Let $\omega \in [0,1]$ and $r > 0$ be the bifurcation parameters. The *sine circle map* is given by

$$\theta_{t+1} = \theta_t + \omega - \frac{r}{2\pi}\sin(2\pi\theta_t), \quad \bmod 1$$

where $t = 0, 1, \ldots$. The *rotation number* ρ is defined by

$$\rho := \lim_{T\to\infty} \frac{1}{T}\sum_{t=1}^{T} |\theta_t - \theta_{t-1}|$$

whenever the limit exists. For $r < 1$ the limit always exists. It can be either a rational or an irrational number. The regions in the $r - \omega$ space where a unique rational number of ρ exists are called *Arnold tongues.* For $r < 1$ there is no overlap of the Arnold tongues. Show that for $r > 1$ the Arnold tongues begin to overlap.

Problem 51. (i) Study the difference equation

$$x_{t+1} = \sqrt{x_t + 2}, \quad t = 0, 1, 2, \ldots$$

with $x_0 = 0$. First find the fixed points.

(ii) Let $x_0 \geq 0$. Discuss the solution of

$$x_{t+1} = \frac{2\sqrt{x_t}}{1+x_t}, \quad t = 0, 1, \ldots.$$

Show that the fixed points are $x_1^* = 0$ and $x_2^* = 1$. The difference equation plays a role in *Landen's transformation.*

Problem 52. Consider the map $f : [0,1] \to [0,1]$

$$f(x) = \sin^2(2\pi x).$$

Find

$$\rho(x) = \lim_{T\to\infty} \sum_{t=0}^{T-1} \delta(x - f^{(t)}(x_0))$$

where $x_0 \in [0,1]$ is the initial value.

Problem 53. Consider the analytic function $f : \mathbb{R} \to \mathbb{R}$, $f(x) = \cos(x)$. The equation $\cos(x) = x$ has one solution, i.e. we have one fixed point for f. Consider the function $f(f(x)) = \cos(\cos(x))$. Does the equation $\cos(\cos(x^*)) = x^*$ admit other solutions besides the one of $\cos(x^*) = x^*$?

Problem 54. Consider the map $f : [0,1] \to [0,1]$ given by

$$f(x) = \begin{cases} 7x/5 & \text{for } x \in [0, 1/2] \\ 14x(1-x)/5 & \text{for } x \in [1/2, 1]. \end{cases}$$

Find a lower as well as an upper limit for the value of the Liapunov exponent and thus show that the map is not chaotic. For the derivative of f in each interval one has

$$\frac{df(x)}{dx} = \begin{cases} 7/5 & \text{for } x \in [0, 1/2] \\ 14(1-2x)/5 & \text{for } x \in [1/2, 1]. \end{cases}$$

Now f is monotonic on both the half intervals. Therefore we have

$$\ln|df(x)/dx| = \ln(7/5) \text{ for } x \in [0, 1/2]$$

and

$$\ln(0) \leq \ln|df(x)/dx| \leq \ln|-14/5| \text{ for } x \in [1/2, 1].$$

On the whole of the interval $[0,1]$ it follows that

$$-\infty \leq \ln|df(x)/dx| \leq \ln(14/5)$$

where $\ln(14/5) \approx 0.103$. The fixed point (besides 0) is given by $x^* = 9/14$.

Problem 55. Let $(s_0, s_1, ..., s_{n-1})^T \in \mathbb{R}^n$, where $n = 2^k$. This vector in $\mathbb{R}^n$ can be associated with a piecewise constant function f defined on $[0, 1)$

$$f(x) = \sum_{j=0}^{2^k-1} s_j \Theta_{[j2^{-k},(j+1)2^{-k})}(x)$$

where $\Theta_{[j2^{-k},(j+1)2^{-k})}(x)$ is the step function with the support $[j2^{-k}, (j+1)2^{-k})$

$$\Theta_{[j2^{-k},(j+1)2^{-k})}(x) := \begin{cases} 1 \, x \in [j2^{-k}, (j+1)2^{-k}) \\ 0 \, x \notin [j2^{-k}, (j+1)2^{-k}) \end{cases}$$

Let $x_{j+1} = 4x_j(1 - x_j)$ with $j = 0, 1, 2, \ldots$ and $x_0 = 1/3$. Then

$$x_0 = \frac{1}{3}, \quad x_1 = \frac{8}{9}, \quad x_2 = \frac{32}{81}, \quad x_3 = \frac{6272}{6561}.$$

Find f and calculate

$$\int_0^1 f(x)dx.$$

For $n = 4$ with $k = 2$ we have the four intervals

$$[0, 1/4), \quad [1/4, 1/2), \quad [1/2, 3/4), \quad [3/4, 1)$$

and

$$\int_0^1 f(x)dx = \left(\frac{1}{3} + \frac{8}{9} + \frac{32}{81} + \frac{6272}{6561} \right) \frac{1}{4}.$$

Problem 56. Given the two sequences of length n

$$\mathbf{S} = (S_0, S_1, \ldots, S_{n-1}), \quad \mathbf{T} = (T_0, T_1, \ldots, T_{n-1})$$

where $S_j, T_j \in \{-1, +1\}$. Implement the "delta function"

$$\delta(\mathbf{S}, \mathbf{T}) = \frac{1}{2^n} \prod_{j=0}^{n-1} (1 + S_j T_j).$$

Let $n > 2$. Calculate the autocorrelation function

$$C_k(\mathbf{S}) = \sum_{j=0}^{n-1-k} S_j S_{j+k}$$

where $k = 1, \ldots, n - 1$.

Problem 57. Let x be a positive integer. Now $\lfloor x \rfloor$ denotes the greatest integer not exceeding x. Let a and b be real numbers with $a > 1$. We define the sequence p_t as

$$p_t = \lfloor (t+1)/a + b \rfloor - \lfloor t/a + b \rfloor, \quad t = 1, 2, \ldots$$

Consequently we have a sequence of 0's and 1's.
(i) Show that the sequence p_t takes its 1's on the set

$$S_1 : \{ t : t = \lfloor (k-b)a \rfloor, k \in \mathbb{R} \}$$

and its 0's on the set

$$S_0 : \{ t : t = \lfloor (\ell + b)c \rfloor, \ell \in \mathbb{N} \}$$

where c is defined as $1/a + 1/c = 1$. (ii) Show that the two sets satisfy $S_0 \cap S_1 = \emptyset$ and $S_0 \cup S_1 = \mathbb{N}$.
(iii) Let $a = 2$ and $b = 1/2$. Write a C++ program using the class `Verylong` that finds the sequence.

Problem 58. Let $\phi = \frac{1}{2}(\sqrt{5} - 1)$. Calculate the sequence

$$x_t = \lfloor (t+1)/\phi \rfloor - \lfloor t/\phi \rfloor, \quad t = 0, 1, 2, \ldots.$$

Problem 59. The *Cantor sequence* is constructed as follows. Given the natural numbers

$$n = 0, 1, 2, 3, 4, 5, 6, 7, 8, 9, \ldots$$

We write them in *ternary notation* as

$$0, 1, 2, 10, 11, 12, 20, 21, 22, 100, \ldots$$

Then the Cantor sequence b_n $(n = 0, 1, 2, \ldots)$ is defined as: if n in ternary has only 0s and 2s, then $b_n = 1$, otherwise we set $b_n = 0$. The first 9 terms of the sequence are given by

$$1, 0, 1, 0, 0, 0, 1, 0, 1, \ldots$$

Write a C++ program that generates this sequence. Is the sequence periodic? Is the sequence chaotic?

Problem 60. Can one find a function f such that

$$x - \frac{f(x)}{df/dx} = 4x(1-x)\,?$$

Chapter 2

Higher-Dimensional Maps and Complex Maps

2.1 Introduction

Let $\mathbf{f} : \mathbb{R}^n \to \mathbb{R}^n$ be a C^1 map and $\mathbf{x}^*$ be a fixed point of $\mathbf{f}$, i.e. $\mathbf{f}(\mathbf{x}^*) = \mathbf{x}^*$. Then $\mathbf{x}^*$ is said to be a hyperbolic fixed point of $\mathbf{f}$ if the Jacobian matrix $D\mathbf{f}(\mathbf{x} = \mathbf{x}^*)$ has no eigenvalues on the unit circle in the complex plane.

Theorem. (Hartman-Grobman) Let U be an open set in $\mathbb{R}^n$. Let $\mathbf{x}^*$ be a hyperbolic fixed point of the diffeomorphism $\mathbf{f} : U \to \mathbb{R}^n$. Then there is a neighbourhood $N \subset U$ of $\mathbf{x}^*$ and a neighbourhood $N' \subseteq \mathbb{R}^n$ containing the origin such that $\mathbf{f}|N$ is topologically conjugate to $D\mathbf{f}(\mathbf{x}^*)|N'$.

In particular we consider maps $\mathbf{f} : \mathbb{R}^n \to \mathbb{R}^n$ with f_j $(j = 1, \ldots, n)$ are smooth functions. The corresponding system of difference equations would be

$$x_{j,t+1} = f_j(x_{1,t}, \ldots, x_{n,t}), \quad j = 1, \ldots, n.$$

The *variational equation* (*linearized equation*) is

$$y_{j,t+1} = \sum_{k=1}^{n} \frac{\partial f_j(\mathbf{x}_t)}{\partial x_k} y_{k,t}, \quad j = 1, \ldots, n.$$

Inverse function theorem. Let $\mathbf{f}$ be a continuously differentiable, vector-valued function mapping an open set $E \subset \mathbb{R}^n$ to $\mathbb{R}^n$. Let $S = \mathbf{f}(E)$. If, for some point $\mathbf{p} \in E$, the Jacobian determinant $\det(J_{\mathbf{f}}(\mathbf{p}))$ is nonzero, then there exists a uniquely determined functional $\mathbf{g}$ and two open sets $X \subset E$ and $Y \subset S$ such that (i) $\mathbf{p} \in X$, $\mathbf{f}(\mathbf{p}) \in Y$, (ii) $Y = \mathbf{f}(X)$, (iii) $\mathbf{f} : X \to Y$ is one-one, (iv) $\mathbf{g}$ is continuously differentiable on Y and

$$\mathbf{g}(\mathbf{f}(\mathbf{x})) = \mathbf{x}$$

for all $\mathbf{x} \in X$.

The *Hopf bifurcation theorem* for maps in the plane $\mathbf{f}_r : \mathbb{R}^2 \to \mathbb{R}^2$, where r is the bifurcation parameter, is as follows.

Theorem. (Hopf bifurcation theorem) Let $\mathbf{f}(r, \mathbf{x})$ be a one-parameter family of maps in the plane satisfying:
a) An isolated fixed point $\mathbf{x}^*(r)$ exists.
b) The map $\mathbf{f}_r$ is C^k ($k \geq 3$) in the neighbourhood of $(\mathbf{x}^*(r_0); r_0)$.
c) The Jacobian matrix $D_{\mathbf{x}}\mathbf{f}(\mathbf{x}^*(r); r)$ possesses a pair of complex, simple eigenvalues

$$\lambda(r) = e^{\alpha(r) + i\omega(r)}$$

and $\bar{\lambda}(r)$, such that the critical value $r = r_0$

$$|\lambda(r_0)| = 1, \quad (\lambda(r_0))^3 \neq 1, \quad (\lambda(r_0))^4 \neq 1, \quad \frac{d|\lambda(r)|}{dr}(r = r_0) > 0.$$

(Existence) Then there exists a real number $\epsilon_0 > 0$ and a C^{k-1} function such that

$$r(\epsilon) = r_0 + r_1\epsilon + r_3\epsilon^3 + O(\epsilon^4)$$

such that for each $\epsilon \in (0, \epsilon_0]$ the map $\mathbf{f}_r$ has an invariant manifold $H(r)$, i.e. $\mathbf{f}(H(r); r) = H(r)$. The manifold $H(r)$ is C^r diffeomorphic to a circle and consists of points at a distance $O(|r|^{1/2})$ of $\mathbf{x}^*(r)$, for $r = r(\epsilon)$.

(Uniqueness) Each compact invariant manifold close to $\mathbf{x}^*(r)$ for $r = r(\epsilon)$ is contained in $H(r) \cup \{0\}$.

(Stability) If $r_3 < 0$ (respectively $r_3 > 0$) then for $r > 0$ (respectively $r > 0$), the fixed point $\mathbf{x}^*(r(\epsilon))$ is stable (respectively unstable) and for $r > 0$ (respectively $r < 0$) the fixed point $\mathbf{x}^*(r(\epsilon))$ is unstable (respectively stable) and the surrounding manifold $H(r(\epsilon))$ is attracting (respectively repelling). When $r_3 < 0$ (respectively $r_3 > 0$) the bifurcation at $r = r(\epsilon)$ is said to be *supercritical* (respectively *subcritical*).

2.2 Two-Dimensional Maps

2.2.1 Solved Problems

Problem 1. Consider the analytic map $\mathbf{f} : \mathbb{R}^2 \to \mathbb{R}^2$

$$f_1(x_1, x_2) = x_2, \qquad f_2(x_1, x_2) = x_1 x_2.$$

The map can be written as a system of difference equations

$$x_{1,t+1} = x_{2,t}, \quad x_{2,t+1} = x_{1,t} x_{2,t} \quad t = 0, 1, 2, \dots$$

Thus we can also write it as a second order difference equation

$$x_{1,t+2} = x_{1,t} x_{1,t+1}.$$

(i) Find $\mathbf{f}^*(dx_1 \wedge dx_2)$.
(ii) Find the fixed points of the map.
(iii) Find the variational equation. Study the stability of the fixed points.
(iv) Find the orbit for $x_{1,0} = 1$, $x_{1,1} = 1/2$.

Solution 1. (i) With

$$df_1 = dx_2, \qquad df_2 = x_2 dx_1 + x_1 dx_2$$

and $dx_1 \wedge dx_1 = dx_2 \wedge dx_2 = 0$, $dx_1 \wedge dx_2 = -dx_2 \wedge dx_1$ we obtain

$$\mathbf{f}^*(dx_1 \wedge dx_2) = -x_2 dx_1 \wedge dx_2$$

which means that

$$\det \begin{pmatrix} \partial f_1/\partial x_1 & \partial f_1/\partial x_2 \\ \partial f_2/\partial x_1 & \partial f_2/\partial x_2 \end{pmatrix} = -x_2.$$

Thus the map is not invertible.
(ii) The fixed points are found from the solutions of the system of equations

$$x_2^* = x_1^*, \qquad x_1^* x_2^* = x_2^*.$$

We find the two fixed points $(0, 0)$ and $(1, 1)$.
(iii) The variational equation is

$$y_{1,t+1} = \frac{\partial f_1(\mathbf{x}_t)}{\partial x_1} y_{1,t} + \frac{\partial f_1(\mathbf{x}_t)}{\partial x_2} y_{2,t}$$

$$y_{2,t+1} = \frac{\partial f_2(\mathbf{x}_t)}{\partial x_1} y_{1,t} + \frac{\partial f_2(\mathbf{x}_t)}{\partial x_2} y_{2,t}.$$

For the derivatives of the functions f_1 and f_2 we have

$$\frac{\partial f_1}{\partial x_1} = 0, \quad \frac{\partial f_1}{\partial x_2} = 1, \quad \frac{\partial f_2}{\partial x_1} = x_2, \quad \frac{\partial f_2}{\partial x_2} = x_1.$$

The fixed point $(0,0)$ provides the system of difference equations

$$y_{1,t+1} = y_{2,t}, \quad y_{2,t+1} = 0$$

and thus the fixed point $(0,0)$ is stable. For the fixed point $(1,1)$ we find

$$y_{1,t+1} = y_{2,t}, \quad y_{2,t+1} = y_{1,t} + y_{2,t}$$

and thus the fixed point $(1,1)$ is unstable.
(iv) For the initial values $x_{1,0} = 1$, $x_{1,1} = 1/2$ we find the orbit

$$x_{1,0} = 1, \; x_{1,1} = \frac{1}{2}, \; x_{1,2} = \frac{1}{2}, \; x_{1,3} = \frac{1}{4}, \; x_{1,4} = \frac{1}{8}, \ldots$$

The orbit tends to the stable fixed point $(0,0)$.

Problem 2. Consider the analytic map $\mathbf{f} : \mathbb{R}^2 \to \mathbb{R}^2$

$$f_1(x_1, x_2) = x_2, \qquad f_2(x_1, x_2) = \frac{x_1}{(x_1 x_2)^2 + 1}.$$

The map can be written as a system of difference equations

$$x_{1,t+1} = x_{2,t}, \quad x_{2,t+1} = \frac{x_{1,t}}{(x_{1,t} x_{2,t})^2 + 1}, \qquad t = 0, 1, 2, \ldots$$

Thus we can also write it as a second order difference equation

$$x_{1,t+2} = \frac{x_{1,t}}{(x_{1,t} x_{1,t+1})^2 + 1}.$$

(i) Find the fixed points of the map.
(ii) Find the variational equation and study the stability of the fixed point.
(iii) Find the orbit for $x_{1,0} = 1$, $x_{1,1} = 1/2$.

Solution 2. (i) From the system of equations

$$x_2^* = (x_1^*)^2, \qquad \frac{x_1^*}{(x_1^* x_2^*)^2 + 1} = x_2^*$$

we obtain the only fixed point $x_1^* = x_2^* = 0$.
(ii) With

$$\frac{\partial f_1(0,0)}{\partial x_1} = 0, \quad \frac{\partial f_1(0,0)}{\partial x_2} = 1, \quad \frac{\partial f_2(0,0)}{\partial x_1} = 1, \quad \frac{\partial f_2(0,0)}{\partial x_2} = 0$$

we obtain the variational equation

$$\begin{pmatrix} y_{1,t+1} \\ y_{2,t+1} \end{pmatrix} = \begin{pmatrix} 0 & 1 \\ 1 & 0 \end{pmatrix} \begin{pmatrix} y_{1,t} \\ y_{2,t} \end{pmatrix}$$

with the solution

$$\begin{pmatrix} y_{1,t} \\ y_{2,t} \end{pmatrix} = \begin{pmatrix} 0 & 1 \\ 1 & 0 \end{pmatrix}^t \begin{pmatrix} y_{1,0} \\ y_{2,0} \end{pmatrix}.$$

(iii) We have

$$x_{1,0} = 1, \quad x_{1,1} = \frac{1}{2}, \quad x_{1,2} = \frac{4}{5}, \quad x_{1,3} = \frac{25}{58}, \quad \dots$$

The sequence tends to the fixed point $(0, 0)$.

Problem 3. Consider the analytic function $\mathbf{f} : \mathbb{R}^2 \to \mathbb{R}^2$

$$f_1(x_1, x_2) = x_1^2 - x_2^2, \quad f_2(x_1, x_2) = 2x_1x_2.$$

The map can be written as a coupled system of difference equations

$$x_{1,t+1} = x_{1,t}^2 - x_{2,t}^2, \qquad x_{2,t+1} = 2x_{1,t}x_{2,t}.$$

(i) Find $\mathbf{f}^*(dx_1 \wedge dx_2)$.
(ii) Show that the fixed points are given by $(0, 0)$ and $(1, 0)$. Study the stability of the fixed points.
(iii) Define r_t and ϕ_t via $x_{1,t} = r_t \cos(\phi_t)$, $x_{2,t} = r_t \sin(\phi_t)$. Show that r_t and ϕ_t satisfy the difference equations

$$r_{t+1} = r_t^2, \qquad \phi_{t+1} = 2\phi_t.$$

(iv) Calculate $x_{1,t+1}^2 + x_{2,t+1}^2$.

Solution 3. (i) With

$$df_1 = 2x_1dx_1 - 2x_2dx_2, \quad df_2 = 2x_2dx_1 + 2x_1dx_2$$

we obtain

$$\mathbf{f}^*(dx_1 \wedge dx_2) = 4(x_1^2 + x_2^2)dx_1 \wedge dx_2.$$

Thus the map is not invertible.
(ii) Solving the system of equations

$$(x_1^*)^2 - (x_2^*)^2 = x_1^*, \qquad 2x_1^*x_2^* = x_2^*$$

we obtain the two fixed points $(0, 0)$ and $(1, 0)$. With

$$\frac{\partial f_1}{\partial x_1} = 2x_1, \quad \frac{\partial f_1}{\partial x_2} = -2x_2, \quad \frac{\partial f_2}{\partial x_1} = 2x_2, \quad \frac{\partial f_2}{\partial x_2} = 2x_1$$

we find for the fixed point $(0,0)$ the variational equation

$$y_{1,t+1} = 0, \qquad y_{2,t+1} = 0.$$

Thus the fixed point $(0,0)$ is stable. For the fixed point $(1,0)$ the variational equation is given by

$$\begin{pmatrix} y_{1,t+1} \\ y_{2,t+1} \end{pmatrix} = \begin{pmatrix} 2 & 0 \\ 2 & 0 \end{pmatrix} \begin{pmatrix} y_{1,t} \\ y_{2,t} \end{pmatrix}.$$

(iii) We have

$$r_{t+1}\cos(\phi_{t+1}) = r_t^2 \cos(2\phi_t)$$
$$r_{t+1}\sin(\phi_{t+1}) = 2r_t^2 \sin(\phi_t)\cos(\phi_t) \equiv r_t^2 \sin(2\phi_t).$$

It follows that $\phi_{t+1} = 2\phi_t$ and therefore $r_{t+1} = r_t^2$. For $r_0 < 1$ we find $\lim_{t\to\infty\infty} r_t = 0$. Thus the fixed point $(0,0)$ is stable. For $r_0 > 1$ we have $\lim_{t\to\infty} r_t = \infty$. Thus the fixed point $(1,0)$ is unstable. For $r_0 = 1$ we obtain $r_t = 1$, i.e. the unit circle is an invariant set.
(iv) We have

$$\begin{aligned}(x_{1,t+1})^2 + (x_{2,t+1})^2 &= (x_{1,t}^2 - x_{2,t}^2)^2 + 4x_{1,t}^2 x_{2,t}^2 \\ &= x_{1,t}^4 + x_{2,t}^4 + 2x_{1,t}^2 x_{2,t}^2 \\ &= ((x_{1,t})^2 + (x_{2,t})^2)^2.\end{aligned}$$

Problem 4. Consider the analytic function $\mathbf{f} : \mathbb{R}^2 \to \mathbb{R}^2$

$$f_1(x_1, x_2) = x_1 + x_2, \quad f_2(x_1, x_2) = x_1 x_2.$$

The map can be written as a coupled system of difference equations

$$x_{1,t+1} = x_{1,t} + x_{2,t}, \qquad x_{2,t+1} = x_{1,t} x_{2,t}.$$

(i) Find $\mathbf{f}^*(dx_1 \wedge dx_2)$.
(ii) Find the fixed points.
(iii) Find the variational equation and solve it.
(iv) Find the fixed points of $\mathbf{f}(\mathbf{f})$.

Solution 4. (i) From

$$df_1 = dx_1 + dx_2, \quad df_2 = x_1 dx_2 + x_2 dx_1$$

we obtain $\mathbf{f}^*(dx_1 \wedge dx_2) = (x_1 - x_2) dx_1 \wedge dx_2$. Thus the map is not invertible.
(ii) From $x_1^* + x_2^* = x_1^*$, $x_1^* x_2^* = x_2^*$ we find $x_2^* = 0$, $x_1^* = c \in \mathbb{R}$.

(iii) We have

$$\frac{\partial f_1}{\partial x_1} = 1, \quad \frac{\partial f_1}{\partial x_2} = 1, \quad \frac{\partial f_2}{\partial x_1} = x_2, \quad \frac{\partial f_2}{\partial x_2} = x_1.$$

Thus the variational equation is

$$y_{1,t+1} = y_{1,t} + y_{2,t}, \qquad y_{2,t+1} = x_{2,t}y_{1,t} + x_{1,t}y_{2,t}$$

Inserting the fixed point $(c, 0)$ we arrive at

$$y_{1,t+1} = y_{1,t} + y_{2,t}, \qquad y_{2,t+1} = cy_{2,t}$$

or in matrix form

$$\begin{pmatrix} y_{1,t+1} \\ y_{2,t+1} \end{pmatrix} = \begin{pmatrix} 1 & 1 \\ 0 & c \end{pmatrix} \begin{pmatrix} y_{1,t} \\ y_{2,t} \end{pmatrix}.$$

The solution is

$$\begin{pmatrix} y_{1,t} \\ y_{2,t} \end{pmatrix} = \begin{pmatrix} 1 & 1 \\ 0 & c \end{pmatrix}^t \begin{pmatrix} y_{1,0} \\ y_{2,0} \end{pmatrix}.$$

Thus leads to the matrix

$$A(x_1, x_2) = \begin{pmatrix} \partial f_1/\partial x_1 & \partial f_1/\partial x_2 \\ \partial f_2/\partial x_1 & \partial f_2/\partial x_2 \end{pmatrix} = \begin{pmatrix} 1 & 1 \\ x_1 & x_2 \end{pmatrix}$$

with the determinant of $A(x_1, x_2)$ equal to $x_1 - x_2$.
(iv) The second iterate is given by

$$f_1(f_1(x_1, x_2), f_2(x_1, x_2)) = x_1 + x_2 + x_1x_2$$

$$f_2(f_1(x_1, x_2), f_2(x_1, x_2)) = (x_1 + x_2)x_1x_2.$$

Thus for the fixed points we have to solve

$$x_1^* + x_2^* + x_1^*x_2^* = x_1^*, \quad (x_1^* + x_2^*)x_1^*x_2^* = x_2^*.$$

Obviously we have the fixed point $(c, 0)$. No new fixed points appear.

Problem 5. Consider the coupled logistic maps $\mathbf{f} : [0,1] \times [0,1] \to [0,1] \times [0,1]$

$$f_1(x_1, x_2) = 4x_2(1 - x_2), \quad f_2(x_1, x_2) = 4x_1(1 - x_1)$$

or written as difference equations

$$x_{1,t+1} = 4x_{2,t}(1 - x_{2,t}), \qquad x_{2,t+1} = 4x_{1,t}(1 - x_{1,t})$$

where $x_{1,0} \in [0,1]$ and $x_{2,0} \in [0,1]$.

(i) Find the fixed points and study their stability.
(ii) Find the *preimage* of $(3/4, 3/4)$.

Solution 5. (i) Solving the system of equations

$$4x_2^*(1-x_2^*) = x_1^*, \qquad 4x_1^*(1-x_1^*) = x_2^*$$

provides the four solutions

$$(0,0), \quad (3/4,3/4), \quad ((\sqrt{5}+5)/8, -(\sqrt{5}-5)/8), \quad (-(\sqrt{5}-5)/8, (\sqrt{5}+5)/8).$$

Note that the set of equations is invariant under $x_1 \mapsto x_2$, $x_2 \mapsto x_1$. The variational equation is given by

$$y_{1,t+1} = \frac{\partial f_1}{\partial x_1} y_{1,t} + \frac{\partial f_1}{\partial x_2} y_{2,t}, \quad y_{2,t+1} = \frac{\partial f_2}{\partial x_1} y_{1,t} + \frac{\partial f_2}{\partial x_2} y_{2,t}.$$

Since $\partial f_1/\partial x_1 = 0$, $\partial f_1/\partial x_2 = 4 - 8x_2$, $\partial f_2/\partial x_1 = 4 - 8x_1$, $\partial f_2/\partial x_2 = 0$ we obtain

$$y_{1,t+1} = (4 - 8x_2)y_{2,t}, \quad y_{2,t+1} = (4 - 8x_1)y_{1,t}.$$

Inserting the fixed points we find that all the fixed points are unstable.
(ii) We have to solve

$$4x_2(1-x_2) = \frac{3}{4}, \quad 4x_1(1-x_1) = \frac{3}{4}.$$

We obtain the four solutions

$$(1/4, 1/4), \quad (1/4, 3/4), \quad (3/4, 1/4), \quad (3/4, 3/4).$$

Does the coupled system show chaotic and hyperchaotic behaviour?

Problem 6. Let $a > 0$ and $b > 0$. Consider the *Hénon map* $\mathbf{f} : \mathbb{R}^2 \to \mathbb{R}^2$

$$f_1(x_1, x_2) = 1 + x_2 - ax_1^2, \quad f_2(x_1, x_2) = bx_1$$

with the differential one-forms

$$df_1 = dx_2 - 2ax_1 dx_1, \qquad df_2 = bdx_1.$$

(i) Let $\omega = dx_1 \wedge dx_2$. Find $\mathbf{f}^*(\omega) = \mathbf{f}^*(dx_1 \wedge dx_2)$. Thus show that the Hénon map is invertible. Find the inverse.
(ii) Let $\alpha = x_1 dx_2 - x_2 dx_1$. Find $\mathbf{f}^*(\alpha)$.
(iii) Consider the *metric tensor field* of $\mathbb{R}^2$

$$g = dx_1 \otimes dx_1 + dx_2 \otimes dx_2.$$

Find $\mathbf{f}^*(g)$.

Solution 6. (i) With $dx_1 \wedge dx_1 = 0$, $dx_2 \wedge dx_2 = 0$, $dx_1 \wedge dx_2 = -dx_2 \wedge dx_1$ we obtain

$$\begin{aligned}\mathbf{f}^*(dx_1 \wedge dx_2) &= df_1(x_1, x_2) \wedge df_2(x_1, x_2) \\ &= (d(1 + x_2 - ax_1^2)) \wedge (d(bx_1)) \\ &= (dx_2 - 2ax_1 dx_1) \wedge (bdx_1) \\ &= -bdx_1 \wedge dx_2.\end{aligned}$$

Thus the determinant of the functional matrix is given by $-b$. Since $b \neq 0$ the Hénon map is invertible. The inverse map is given by

$$x_{1,t} = \frac{1}{b} x_{2,t+1}, \qquad x_{2,t} = x_{1,t+1} - 1 + \frac{a}{b} x_{2,t+1}^2.$$

(ii) We have

$$\begin{aligned}\mathbf{f}^*(\alpha) &= f_1 df_2 - f_2 df_1 \\ &= (1 + x_2 - ax_1^2) bdx_1 - bx_1(dx_2 - 2ax_1 dx_1) \\ &= b(1 + x_2 + ax_1^2) dx_1 - bx_1 dx_2.\end{aligned}$$

(iii) We have

$$\mathbf{f}^*(g) = (b^2 + 4a^2 x_1^2) dx_1 \otimes dx_1 - 2ax_1 dx_1 \otimes dx_2 - 2ax_1 dx_2 \otimes dx_1 + dx_2 \otimes dx_2$$

or written in matrix form

$$\begin{pmatrix} b^2 + 4a^2 x_1^2 & -2ax_1 \\ -2ax_1 & 1 \end{pmatrix}.$$

The determinant of this matrix is equal to b^2.

Problem 7. Let $a > 0$, $b > 0$ be the bifurcation parameters. The *Hénon map* $\mathbf{f}_{a,b} : \mathbb{R}^2 \to \mathbb{R}^2$ is defined by

$$f_1(x_1, x_2) = 1 + x_2 - ax_1^2 \qquad f_2(x_1, x_2) = bx_1$$

or written as system of difference equations

$$x_{1,t+1} = 1.0 + x_{2,t} - ax_{1,t}^2, \quad x_{2,t+1} = bx_{1,t}, \quad t = 0, 1, \ldots$$

(i) Find the fixed points.
(ii) Show that for $a = 1.4$ and $b = 0.3$ the fixed points are unstable.
(iii) Find the periodic points of period 2 of $\mathbf{f}_{a,b}$ for $a = 1.4$ and $b = 0.3$. Show that the periodic points are unstable.

Solution 7. (i) The solution of the equations

$$1 - x_2^* - ax_1^* = x_1^*, \qquad bx_1^* = x_2^*$$

provide the two fixed points

$$x_1^* = \frac{1}{2}[-(1-b) + \sqrt{(1-b)^2 + 4a}], \qquad x_2^* = bx_1^*$$

$$x_1^* = \frac{1}{2}[-(1-b) - \sqrt{(1-b)^2 + 4a}], \qquad x_2^* = bx_1^*.$$

Thus we have real roots if

$$a \geq -\frac{(1-b)^2}{4} := a_0(b).$$

One has $x_1 \geq x_2$, with the equality occurring for $a = a_0(b)$. The fixed points are real for $a > (1-b)^2/4$. For $a = 1.4$ and $b = 0.3$ the fixed points are real.
(ii) The stability of the fixed points are determined by the *Jacobian matrix*, defined as

$$D\mathbf{f}_{a,b}(x_1, x_2) := \begin{pmatrix} \dfrac{\partial f_1}{\partial x_1}(x_1, x_2) & \dfrac{\partial f_1}{\partial x_2}(x_1, x_2) \\ \dfrac{\partial f_2}{\partial x_1}(x_1, x_2) & \dfrac{\partial f_2}{\partial x_2}(x_1, x_2) \end{pmatrix} = \begin{pmatrix} -2ax_1 & 1 \\ b & 0 \end{pmatrix}.$$

Inserting the fixed points with $a = 1.4$ and $b = 0.3$ shows that the fixed points are unstable.
(iii) Periodic points (x_1, x_2) of period 2 satisfy the simultaneous equations

$$x_1^* = f_1^{(2)}(x_1^*, x_2^*), \qquad x_2^* = f_2^{(2)}(x_1^*, x_2^*)$$

where

$$f_1^{(2)}(x_1, x_2) = f_1(f_1(x_1, x_2), f_2(x_1, x_2)) = 1 + bx_1 - a(1 + x_2 - ax_1^2)^2$$
$$f_2^{(2)}(x_1, x_2) = f_2(f_1(x_1, x_2), f_2(x_1, x_2)) = b(1 + x_2 - ax_1^2).$$

The period 2 points therefore satisfy the simultaneous equations

$$x_1 = 1 + bx_1 - a(1 + x_2 - ax_1^2)^2, \qquad x_2 = b(1 + x_2 - ax_1^2).$$

Consequently

$$x_2 = \frac{b(1 - ax_1^2)}{1 - b}$$

and we obtain the quartic equation

$$a^3x_1^4 - 2a^2x_1^2 + (1-b)^3x_1 + a - (1-b)^2 = 0.$$

This equation also yields the values of the fixed points found in (ii). Therefore, the left-hand side can be factored out. We obtain the quadratic equation

$$a^2 x_1^2 - a(1-b)x_1 + (1-b)^2 - a = 0$$

with solutions

$$\frac{(1-b) \pm \sqrt{4a - 3(1-b)^2}}{2a}.$$

The corresponding values of x_2 are found as

$$\frac{b\left((1-b) \mp \sqrt{4a - 3(1-b)^2}\right)}{2a}.$$

Therefore, these periodic points exist for $a = 1.4$ and $b = 0.3$ and they are unstable.

Problem 8. Consider the analytic function $\mathbf{f} : \mathbb{R}^2 \to \mathbb{R}^2$

$$f_1(x_1, x_2) = \sinh(x_2), \quad f_2(x_1, x_2) = \sinh(x_1).$$

(i) Let $\omega = dx_1 \wedge dx_2$. Find $\mathbf{f}^*(\omega)$.
(ii) Show that this function admits the (only) fixed point $(0,0)$. Find the functional matrix at the fixed point

$$\left.\begin{pmatrix} \partial f_1/\partial x_1 & \partial f_1/\partial x_2 \\ \partial f_2/\partial x_1 & \partial f_2/\partial x_2 \end{pmatrix}\right|_{(0,0)}.$$

(iii) Consider the analytic function $\mathbf{g} : \mathbb{R}^2 \to \mathbb{R}^2$

$$g_1(x_1, x_2) = \sinh(x_1), \quad g_2(x_1, x_2) = -\sinh(x_2).$$

The function $\mathbf{g}$ admits the (only) fixed point $(0,0)$. Find the functional matrix at the fixed point

$$\left.\begin{pmatrix} \partial g_1/\partial x_1 & \partial g_1/\partial x_2 \\ \partial g_2/\partial x_1 & \partial g_2/\partial x_2 \end{pmatrix}\right|_{(0,0)}.$$

(iv) Multiply the two matrices found in (i) and (ii).
(v) Find the composite function $\mathbf{h} : \mathbb{R}^2 \to \mathbb{R}^2$

$$\mathbf{h}(\mathbf{x}) = (\mathbf{f} \circ \mathbf{g})(\mathbf{x}) = \mathbf{f}(\mathbf{g}(\mathbf{x})).$$

Show that this function also admits the fixed point $(0,0)$. Find the functional matrix at this fixed point

$$\left.\begin{pmatrix} \partial h_1/\partial x_1 & \partial h_1/\partial x_2 \\ \partial h_2/\partial x_1 & \partial h_2/\partial x_2 \end{pmatrix}\right|_{(0,0)}.$$

Compare this matrix with the matrix found in (iii).

Solution 8. (i) Since $d\sinh(x)/dx = \cosh(x)$ we obtain

$$\mathbf{f}^*(\omega) = \cosh(x_1)\cosh(x_2) dx_1 \wedge dx_2.$$

Hence the map $\mathbf{f}$ is invertible.
(ii) Since $\sinh(0) = 0$ we find from $\sinh(x_2) = x_1$, $\sinh(x_1) = x_2$ that $x_1 = 0$, $x_2 = 0$. Since $d\sinh(x)/dx = \cosh(x)$ the functional matrix at $(0,0)$ is

$$\begin{pmatrix} 0 & 1 \\ 1 & 0 \end{pmatrix}$$

The map is invertible.
(iii) Since $\sinh(0) = 0$ we find from $\sinh(x_1) = x_1$, $-\sinh(x_2) = x_2$ that $x_1 = 0$, $x_2 = 0$. Since $d\sinh(x)/dx = \cosh(x)$ the functional matrix at $(0,0)$ is

$$\begin{pmatrix} 1 & 0 \\ 0 & -1 \end{pmatrix}.$$

(iv) Multiplication of the two matrices yields

$$\begin{pmatrix} 0 & 1 \\ 1 & 0 \end{pmatrix}\begin{pmatrix} 1 & 0 \\ 0 & -1 \end{pmatrix} = \begin{pmatrix} 0 & -1 \\ 1 & 0 \end{pmatrix}.$$

(v) We have

$$\mathbf{h}(\mathbf{x}) = (f_1(g_1, g_2), f_2(g_1, g_2)).$$

Thus

$$h_1(x_1, x_2) = -\sinh(\sinh(x_2)), \quad h_2(x_1, x_2) = \sinh(\sinh(x_1))$$

which admits the fixed point $(0,0)$. For the functional matrix of $\mathbf{h}$ at $(0,0)$ we find

$$\begin{pmatrix} 0 & -1 \\ 1 & 0 \end{pmatrix}.$$

This is the matrix from (iv).

Problem 9. *Baker's transformation* $\mathbf{f} : [0,1] \times [0,1] \to [0,1] \times [0,1]$ is given by

$$\mathbf{f}(x_1, x_2) = \begin{cases} (2x_1, x_2/2) & 0 \le x_1 < 1/2 \\ (2x_1 - 1, x_2/2 + 1/2) & 1/2 < x_1 \le 1 \end{cases}$$

The horizontal direction is stretched by a factor 2, the vertical direction is contracted by a factor 1/2. The Baker's transformation is a completely chaotic (Bernoulli) area preserving map of the unit square onto itself.

(i) Find $\mathbf{f}(1/3, 2/3)$ and $\mathbf{f}(\mathbf{f}(1/3, 2/3))$.
(ii) Find the fixed points of $\mathbf{f}$.
(iii) Find the inverse of $\mathbf{f}$.
(iv) Find $\mathbf{f}(1/2, 1/2)$, $\mathbf{f}(\mathbf{f}(1/2, 1/2))$ etc. Does this sequence converge?
(v) Find the Frobenius-Perron operator for $\mathbf{f}$.
(vi) Find an explicit expression for the n-th iterate $\mathbf{f}^{(n)}$ of $\mathbf{f}$ in terms of certain permutations σ_n on sets of integers.

Solution 9. (i) We obtain

$$\mathbf{f}(1/3, 2/3) = (2/3, 1/3), \quad \mathbf{f}(2/3, 1/3) = (1/3, 2/3)$$

i.e. we have a periodic orbit.
(ii) Obviously the fixed points are

$$(x_1^*, x_2^*) = (0, 0), \qquad (x_1^*, x_2^*) = (1, 1).$$

(iii) The inverse map is

$$\mathbf{f}^{-1}(x_1, x_2) = \begin{cases} (x_1/2, 2x_2) & \text{if } 0 \le x_2 < 1/2 \\ ((x_1+1)/2, 2x_2 - 1) & \text{if } 1/2 \le x_2 < 1. \end{cases}$$

(iv) We obtain

$$\mathbf{f}(1/2, 1/2) = (1, 1/4), \quad \mathbf{f}(\mathbf{f}(1/2, 1/2)) = (1, 3/4).$$

The sequence converges to the fixed point $(1, 1)$.
(v) We have

$$P_{\mathbf{f}}\rho(x_1, x_2) = \begin{cases} \rho(x_1/2, 2x_2) & 0 \le x_2 < 1/2 \\ \rho(x_1/2, 2x_2 - 1) & 1/2 < x_2) \le 1 \end{cases}$$

(vi) For each positive integer n, the function σ_n maps each integer m in the set $\{0, 1, \ldots, 2^{n-1}\}$ onto the integer (in the same set) whose binary representation as a string of n 0's and 1's is obtained by reversing the corresponding binary representation of m. If

$$m = \sum_{k=0}^{n-1} \epsilon_k 2^k$$

where $\epsilon_k \in \{0, 1\}$ for $k = 0, 1, \ldots, n-1$, then

$$\sigma_n(m) = \sum_{k=0}^{n-1} \epsilon_k 2^{n-1-k}.$$

Thus $\sigma_n(1) = 2^{n-1}$ for any n, $\sigma_n(2) = 2^{n-2}$ for $n \geq 2$, $\sigma_n(3) = 3 \cdot 2^{n-2}$ for $n \geq 2$, $\sigma_n(0) = 0$ and $\sigma_n(2^n - 1) = 2^n - 1$ for any n, etc. For any m in the set $\{0, 1, \ldots, 2^n - 1\}$ we have

$$\sigma_{n+1}(2m) = \sigma_n(m), \qquad \sigma_{n+1}(2m+1) = \sigma_n(m) + 2^n.$$

Then an induction on n yields that for each positive integer n, any integer m such that $0 \leq m \leq 2^n - 1$, and every point (x_1, x_2) in $[0,1] \times [0,1]$ with $m/2^n \leq x_1 < (m+1)/2^n$ we have

$$f^{(n)}(x_1, x_2) = \left(2^n x_1 - m, \frac{1}{2^n}x_2 + \frac{1}{2^n}\sigma_n(m)\right).$$

Problem 10. The phase coupled logistic map $\mathbf{f}_{r,\epsilon} : \mathbb{R}^2 \to \mathbb{R}^2$ is defined by

$$f_{1,r,\epsilon}(x_1, x_2) = rx_1(1 - x_1) + \epsilon(x_2 - x_1)$$
$$f_{2,r,\epsilon}(x_1, x_2) = rx_2(1 - x_2) + \epsilon(x_1 - x_2)$$

with $r > 0$ and $\epsilon \geq 0$ are bifurcation parameters. The system admits the fixed points (among others)

$$(0,0), \qquad ((r-1)/r, (r-1)/r).$$

Find the variational equation and analyse the stability of these fixed points.

Solution 10. The variational equation is defined by

$$y_{1,t+1} = \frac{\partial f_1}{\partial x_1}(x_1 = x_{1,t}, x_2 = x_{2,t})y_{1,t} + \frac{\partial f_1}{\partial x_2}(x_1 = x_{1,t}, x_2 = x_{2,t})y_{2,t}$$

$$y_{2,t+1} = \frac{\partial f_2}{\partial x_1}(x_1 = x_{1,t}, x_2 = x_{2,t})y_{1,t} + \frac{\partial f_2}{\partial x_2}(x_1 = x_{1,t}, x_2 = x_{2,t})y_{2,t}.$$

The stability of the fixed points are determined by the eigenvalues of the Jacobian matrix, i.e.

$$D\mathbf{f}(x_1, x_2) = \begin{pmatrix} \frac{\partial f_1}{\partial x_1}(x_1, x_2) & \frac{\partial f_1}{\partial x_2}(x_1, x_2) \\ \frac{\partial f_2}{\partial x_1}(x_1, x_2) & \frac{\partial f_2}{\partial x_2}(x_1, x_2) \end{pmatrix} = \begin{pmatrix} r(1 - 2x_1) - \epsilon & \epsilon \\ \epsilon & r(1 - 2x_2) - \epsilon \end{pmatrix}.$$

(a) $x_1^* = x_2^* = 0$: We obtain

$$\lambda^2 - 2(r - \epsilon)\lambda + r(r - 2\epsilon) = 0$$

with (real) roots

$$\lambda_+ = r, \quad \lambda_- = r - 2\epsilon.$$

Hence $|\lambda_+| < 1$ if $0 < r < 1$ and $|\lambda_-| < 1$ if $0 < r < 1 + 2\epsilon$. Therefore this fixed point is an attractor for $0 < r < \min\{1, 1+2\epsilon\}$, a saddle point for $\min\{1, 1+2\epsilon\} < r < \max\{1, 1+2\epsilon\}$ and a repellor for $r > \max\{1, 1+2\epsilon\}$.

(b) $x_1^* = x_2^* = \frac{r-1}{r}$: We obtain

$$\lambda^2 - 2(2 - r - \epsilon)\lambda + (2 - r)(2 - r - 2\epsilon) = 0,$$

with (real) roots

$$\lambda_+ = 2 - r, \quad \lambda_- = 2 - r - 2\epsilon.$$

Hence $|\lambda_+| < 1$ if $1 < r < 3$ and $|\lambda_-| < 1$ if $1 - 2\epsilon < r < 3 - 2\epsilon$. Therefore this fixed point is an attractor if $\max\{1, 1 - 2\epsilon\} < r < \min\{3, 3 - 2\epsilon\}$, a saddle point if $\min\{1, 1 - 2\epsilon\} < r < \max\{1, 1 - 2\epsilon\}$ or $\min\{3, 3 - 2\epsilon\} < r < \max\{3, 3 - 2\epsilon\}$ and a repellor if $r > \max\{3, 3 - 2\epsilon\}$.

Problem 11. Consider the Hénon area-preserving map ($\mathbb{R}^2 \to \mathbb{R}^2$)

$$x_{1,t+1} = x_{1,t}\cos(\alpha) - x_{2,t}\sin(\alpha) + x_{1,t}^2\sin(\alpha)$$
$$x_{2,t+1} = x_{1,t}\sin(\alpha) + x_{2,t}\cos(\alpha) - x_{1,t}^2\cos(\alpha).$$

Show that the two fixed points lie on the line $x_2 = x_1\tan(\alpha/2)$.

Solution 11. The fixed points are the solutions of the equations

$$x_1^* = x_1^*\cos(\alpha) - x_2^*\sin(\alpha) + x_1^{*2}\sin(\alpha)$$
$$x_2^* = x_1^*\sin(\alpha) + x_2^*\cos(\alpha) - x_1^{*2}\cos(\alpha).$$

Multiplying the first equation with $\cos(\alpha)$ and multiplying the second equation with $\sin(\alpha)$ and adding yields

$$x_1^*\cos(\alpha) + x_2^*\sin(\alpha) = x_1^*.$$

If $\alpha \neq n\pi$ with $n \in \mathbb{Z}$ we have

$$x_2^* = x_1^*\frac{1 - \cos(\alpha)}{\sin(\alpha)} = x_1^*\tan(\alpha/2).$$

Problem 12. Let $1/2 \le r < 1$ be the bifurcation parameter. Consider the two-dimensional map $\mathbf{f}_r : [0,1]^2 \to [0,1]^2$ given by

$$\mathbf{f}_r(x_1, x_2) = \begin{cases} (2x_1, rx_2) & \text{if } 0 \le x_1 \le 1/2 \\ (2x_1 - 1, rx_2 + (1 - r)) & \text{if } 1/2 \le x_1 \le 1 \end{cases}$$

Discuss the behaviour of the map for $r \geq 1/2$.

Solution 12. The map $\mathbf{f}_r$ stretches horizontally by a factor of 2 and compresses vertically by a factor of r. For $r = 1/2$ we have the classical Baker's transformation. For $r > 1/2$, the map $\mathbf{f}_r$ is not invertible. The map is called *fat Baker's transformation.*

Problem 13. Consider the map $\mathbf{f} : \mathbb{R}^2 \to \mathbb{R}^2$

$$\mathbf{f}(x_1, x_2) = ((4/\pi)\arctan(x_1), x_2/2)$$

i.e. $f_1(x_1, x_2) = \frac{4}{\pi}\arctan(x_1)$, $f_2(x_1, x_2) = x_2/2$.
(i) Find the fixed points. Discuss.
(ii) Find the stable manifold of $(0,0)$. Find the unstable manifold of $(0,0)$. Find the basins of attraction.

Solution 13. (i) Since

$$\arctan(0) = 0, \quad \arctan(1) = \pi/4, \quad \arctan(-1) = -\pi/4$$

we obtain the three fixed points $(0,0)$, $(-1,0)$, $(1,0)$. The fixed points $(1,0)$, $(-1,0)$ are fixed point attractors and the fixed point $(0,0)$ is a fixed point saddle.
(ii) The stable manifold of $(0,0)$ is the x_2-axis. The unstable manifold of $(0,0)$ is the set

$$\{\, (x_1, x_2) \,:\, -1 < x_1 < 1 \text{ and } x_2 = 0 \,\}.$$

The orbits of all points in the left half plane are attracted to $(-1,0)$, and those points in the right half plane are attracted to $(1,0)$.

Problem 14. The *trace map* is given by

$$x_{t+1} = 1 + 4x_{t-1}^2(x_t - 1), \qquad t = 1, 2, \ldots \tag{1}$$

with the initial values x_0, x_1 given. Show that

$$x_{t+1} = 2x_t^2 - 1, \quad t = 0, 1, 2, \ldots \tag{2}$$

is an *invariant* of the trace map.

Definition. Equation (2) is an *invariant* of (1) if (2) is satisfied for the pair (x_t, x_{t+1}) then (1) implies that (x_{t+1}, x_{t+2}) also satisfies (2).

Solution 14. Let us first give an example using numbers. The point $(x_0, x_1) = (1/3, -7/9)$ satisfies (2) with $t = 0$. If $n = 1$ the trace map takes the form

$$x_2 = 1 + 4x_0^2(x_1 - 1). \tag{3}$$

Inserting $x_0 = 1/3$ and $x_1 = -7/9$ into this equation yields $x_2 = 17/81$. The conclusion is that $(x_1, x_2) = (-7/9, 17/81)$ satisfies (2) for $t = 1$. Let us now give the general proof. From the trace map we obtain

$$x_{t+2} = 1 + 4x_t^2(x_{t+1} - 1). \qquad (4)$$

Inserting the invariant (2) into (4) yields

$$x_{t+2} = 1 + 4x_t^2(2x_t^2 - 2). \qquad (5)$$

From the invariant (2) we also obtain

$$x_{t+2} = 2x_{t+1}^2 - 1. \qquad (6)$$

Inserting (2) into (6) yields

$$x_{t+2} = 2(4x_t^4 + 1 - 4x_t^2) - 1 = 1 + 4x_t^2(2x_t^2 - 2). \qquad (7)$$

Thus the right hand side of (7) and (5) coincide and therefore (2) is an invariant of (1).

Problem 15. Consider the system of difference equations

$$x_{1,t+1} = x_{2,t} + rx_{1,t}(1 - x_{1,t}^2 - x_{2,t}^2)$$

$$x_{2,t+1} = -x_{1,t} + rx_{2,t}(1 - x_{1,t}^2 - x_{2,t}^2)$$

where r is a real bifurcation parameter and $t = 0, 1, 2, \ldots$.
(i) Show that for $r = 0$ the system admits a *first integral.*
(ii) Show that for $r \neq 0$ the system admits the one-dimensional *integral manifold*

$$x_{1,t}^2 + x_{2,t}^2 - 1 = 0.$$

Solution 15. (i) For $r = 0$ we have

$$x_{1,t+1}^2 + x_{2,t+1}^2 = x_{2,t}^2 + x_{1,t}^2.$$

(ii) For $r \neq 0$ straightforward calculation shows

$$x_{1,t+1}^2 + x_{2,t+1}^2 - 1 = x_{1,t}^2 + x_{2,t}^2 - 1 + r^2(x_{1,t}^2 + x_{2,t}^2)(1 - x_{1,t}^2 - x_{2,t}^2).$$

Thus the system admits the one-dimensional integral manifold $x_{1,t}^2 + x_{2,t}^2 = 1$.

Problem 16. A one-dimensional map f is called an *invariant* of a two-dimensional map g if

$$g(x, f(x)) = f(f(x)).$$

Let

$$f(x) = 2x^2 - 1.$$

Show that f is an invariant for

$$g(x, y) = y - 2x^2 + 2y^2 + d(1 + y - 2x^2).$$

Solution 16. We have

$$f(f(x)) = 2(2x^2 - 1)^2 - 1.$$

On the other hand we have

$$g(x, f(x)) = 2x^2 - 1 - 2x^2 + 2(2x^2 - 1)^2 + d(1 + 2x^2 - 1 - 2x^2).$$

Thus

$$g(x, f(x)) = -1 + 2(2x^2 - 1)^2.$$

Consequently $g(x, f(x)) = f(f(x))$.

Problem 17. (i) Given the analytic maps $\mathbf{h} : \mathbb{R}^2 \to \mathbb{R}^2$, $\mathbf{g} : \mathbb{R}^2 \to \mathbb{R}^2$. Find

$$\mathbf{f}(\mathbf{x}) = (\mathbf{g} \circ \mathbf{h})(\mathbf{x}) = \mathbf{g}(\mathbf{h}(\mathbf{x})).$$

(ii) Apply it to

$$h_1(x_1, x_2) = x_1 x_2, \quad h_2(x_1, x_2) = x_1 + x_2$$

$$g_1(x_1, x_2) = x_1 + x_2, \quad g_2(x_1, x_2) = x_1 x_2.$$

Note that $\mathbf{h}$ and $\mathbf{g}$ are invariant under $x_1 \leftrightarrow x_2$ and both admit the fixed point $(0, 0)$. Is $\mathbf{f}$ invariant under $x_1 \leftrightarrow x_2$? Does $\mathbf{f}$ admit the fixed point $(0, 0)$.

Solution 17. (i) We find

$$f_1(x_1, x_2) = g_1(h_1(x_1, x_2), h_2(x_1, x_2))$$
$$f_2(x_1, x_2) = g_2(h_1(x_1, x_2), h_2(x_1, x_2)).$$

(ii) For the given example we have

$$f_1(x_1, x_2) = g_1(h_1(x_1, x_2), h_2(x_1, x_2)) = x_1 x_2 + x_1 + x_2$$
$$f_2(x_1, x_2) = g_2(h_1(x_1, x_2), h_2(x_1, x_2)) = (x_1 x_2)(x_1 + x_2).$$

Obviously $\mathbf{f}$ admits the fixed point $(0, 0)$ and the invariance under $x_1 \leftrightarrow x_2$ is also preserved.

Problem 18. Let $r \in \mathbb{R}$. Consider the map $\mathbf{f} : \mathbb{R}^2 \to \mathbb{R}^2$

$$f_1(x_1, x_2) = x_1 + x_2, \qquad f_2(x_1, x_2) = x_1 - r\sin(x_1 + x_2)$$

or written as difference equation

$$x_{1,t+1} = x_{1,t} + x_{2,t}, \qquad x_{2,t+1} = x_{1,t} - r\sin(x_{1,t} + x_{2,t}).$$

Show that the map $\mathbf{f}$ can be expressed as a product of two shear maps $\mathbf{g}$ and $\mathbf{h}$ with $\mathbf{f}(\mathbf{x}) = (\mathbf{g} \circ \mathbf{h})(\mathbf{x}) = \mathbf{g}(\mathbf{h}(\mathbf{x}))$. Study the case $r = 0$.

Solution 18. We have

$$g_1(x_1, x_2) = x_1, \qquad g_2(x_1, x_2) = x_1 - x_2 - r\sin(x_1)$$

and

$$h_1(x_1, x_2) = x_1 + x_2, \qquad h_2(x_1, x_2) = x_2.$$

For $r = 0$ we have the matrix identity

$$\begin{pmatrix} 1 & 1 \\ 1 & 0 \end{pmatrix} = \begin{pmatrix} 1 & 0 \\ 1 & -1 \end{pmatrix} \begin{pmatrix} 1 & 1 \\ 0 & 1 \end{pmatrix}.$$

Note that the matrices on the right-hand side are nonnormal.

Problem 19. (i) Give an interpretation of the *Arnold cat map*

$$\begin{pmatrix} x_1 \\ x_2 \end{pmatrix} \mapsto \begin{pmatrix} 1 & 1 \\ 1 & 2 \end{pmatrix} \begin{pmatrix} x_1 \\ x_2 \end{pmatrix} \mod 1.$$

(ii) Find the inverse of the *Arnold cat map.* Note that the determinant of the matrix is $+1$ and thus the determinant of the inverse matrix is also $+1$.

Solution 19. (i) Since

$$\begin{pmatrix} 1 & 0 \\ 1 & 1 \end{pmatrix} \begin{pmatrix} 1 & 1 \\ 0 & 1 \end{pmatrix} = \begin{pmatrix} 1 & 1 \\ 1 & 2 \end{pmatrix}$$

we find that the Arnold cat map is the composition of a shear in the x_1-direction with factor 1 followed by a shear in the x_2-direction. The three matrices are elements of the Lie group $SL(2, \mathbb{R})$. Owing to mod 1 all points $(x_1, x_2) \in \mathbb{R}^2$ are mapped into the unit square $[0, 1)^2$.
(ii) We find

$$\begin{pmatrix} x_1 \\ x_2 \end{pmatrix} \mapsto \begin{pmatrix} 2 & -1 \\ -1 & 1 \end{pmatrix} \begin{pmatrix} x_1 \\ x_2 \end{pmatrix} \mod 1.$$

Problem 20. The *Arnold cat map* is defined as follows: $\Omega = [0,1)^2$,

$$T(x_1, x_2) = (x_1 + x_2, x_1 + 2x_2).$$

In matrix form we have

$$\begin{pmatrix} x_1 \\ x_2 \end{pmatrix} \mapsto \begin{pmatrix} 1 & 1 \\ 1 & 2 \end{pmatrix} \begin{pmatrix} x_1 \\ x_2 \end{pmatrix} \quad \text{mod } 1.$$

Thus T maps Ω $1-1$ onto itself.
(i) Show that the map preserves Lebesgue measure.
(ii) Show that T is invertible. Show that the entire sequence can be recovered from one term.
(iii) Find the n-th iterate of T.

Solution 20. (i) The Jacobian determinant of the matrix of the map is equal to one, i.e.

$$\det \begin{pmatrix} 1 & 1 \\ 1 & 2 \end{pmatrix} = 1.$$

Thus the map T preserves Lebesgue measure.
(ii) The inverse of the matrix is given by

$$\begin{pmatrix} 2 & -1 \\ -1 & 1 \end{pmatrix}.$$

Note that these matrices are elements of the Lie group $SL(2, \mathbb{R})$. Since T is invertible, the entire sequence can be recovered from one term.
(iii) The n-th iterate is given by

$$T^{(n)}(x_1, x_2) = (a_{2n-2}x_1 + a_{2n-1}x_2, a_{2n-1}x_1 + a_{2n}x_2)$$

where the a_n are the *Fibonacci numbers* given by $a_0 = a_1 = 1$ and $a_{n+1} = a_n + a_{n-1}$ for $n \geq 1$. Note that

$$a_{2n-2} + 2a_{2n-1} + a_{2n} = a_{2n} + a_{2n+1}.$$

Consider the Hilbert space $L_2([0,1) \times [0,1))$ of square integrable functions. Let f and g be two elements of this Hilbert space defined by

$$f(x_1, x_2) := \exp(2\pi i(px_1 + qx_2)) \equiv \exp(2\pi i p x_1)\exp(2\pi i q x_2)$$
$$g(x_1, x_2) := \exp(2\pi i(rx_1 + sx_2)) \equiv \exp(2\pi i r x_1)\exp(2\pi i s x_2)$$

where $p, q, r, s \in \mathbb{Z}$. Since ($k \in \mathbb{Z}$)

$$\int_0^1 \exp(2\pi i k x)dx \equiv \int_0^1 (\cos(2\pi k x) + i\sin(2\pi k x))dx = 0$$

unless $k = 0$ it follows that

$$\int_0^1 \int_0^1 f(x_1, x_2) g(T^{(n)}(x_1, x_2)) dx_1 dx_2 = 0$$

unless

$$ra_{2n-2} + sa_{2n-1} + p = 0, \quad ra_{2n-1} + sa_{2n} + q = 0.$$

Problem 21. In matrix form the *Arnold cat map* is given by

$$T = \begin{pmatrix} 1 & 1 \\ 1 & 2 \end{pmatrix} \begin{pmatrix} x_1 \\ x_2 \end{pmatrix} \mod 1.$$

(i) Given the recursion for the Fibonacci numbers

$$F_n = F_{n-1} + F_{n-2}, \qquad F_1 = F_2 = 1$$

find $T^{(n)}$ by induction.
(ii) Show that the determinant of $T^{(n)}$ is equals 1.
(iii) Find the eigenvalues of $T^{(n)}$.
(iv) Show that fixed points of $T^{(n)}$ correspond to orbits of period length n and any divisors of n for $n = 2$.

Solution 21. (i) The nth iterate of the Arnold cat map is given by

$$T^{(n)} = \begin{pmatrix} F_{2n-1} & F_{2n} \\ F_{2n} & F_{2n+1} \end{pmatrix}$$

which inherits its area preservation from that of T.
(ii) The determinant of $T^{(n)}$ is given by

$$\det(T^{(n)}) = F_{2n-1}F_{2n+1} - F_{2n}^2.$$

Using the Fibonacci sequence twice we have

$$F_{2n-1}F_{2n+1} - F_{2n}^2 = F_{2n-3}F_{2n-1} - F_{2n-2}^2.$$

By induction we find that

$$\det(T^{(n)}) = F_1 F_3 - F_2^2 = 1.$$

(iii) The two eigenvalues of T are

$$\lambda_1 = \frac{3}{2} \pm \frac{\sqrt{5}}{2}.$$

Since the n-th iterate $T^{(n)}$ is a symmetric matrix over $\mathbb{R}$, it has only real eigenvalues. We find $\lambda_1^{(n)} = \lambda_1^n > 1$ and $\lambda_2^{(n)} = \lambda_2^n < 1$.

(iv) For $n = 2$ we have

$$T^{(2)} = \begin{pmatrix} 2 & 3 \\ 3 & 5 \end{pmatrix}.$$

To find the fixed points of period length 2, one has to solve the two coupled equations

$$2x_1 + 3x_2 = x_1 \bmod 1, \qquad 3x_1 + 5x_2 = x_2 \bmod 1.$$

Apart from the solution $x_1 = x_2 = 0$ (which has period length 1), eliminating x_2 results in $5x_1 = 0 \bmod 1$, that is, $x_1 = k/5$, where k is an integer. The only allowed values for $0 < x_1 < 1$ are $k = 1, 2, 3, 4$, each of which, in fact, yields a solution. The corresponding values of x_2 are $x_2 = k'/5$, with $k' = -2k \bmod 5$. These four periodic points of period length 2 form two orbits, namely

$$\left(\frac{2}{5}, \frac{1}{5}\right) \to \left(\frac{3}{5}, \frac{4}{5}\right)$$

and

$$\left(\frac{1}{5}, \frac{3}{5}\right) \to \left(\frac{4}{5}, \frac{2}{5}\right).$$

Problem 22. The family of invertible transformations $\mathbf{f}_{\alpha,p}$, acts on the torus $\mathbb{T}^2$, where $\mathbb{T}$ denotes the circle $[0, 2\pi]$, $\alpha \in \mathbb{T}$, and p is any nonzero integer. It is defined by

$$\mathbf{f}_{\alpha,p}(x_1, x_2) = (x_1 + \alpha, px_1 + x_2).$$

(i) Show that the normalized Lebesgue measure

$$d\mu(x_1, x_2) = \frac{1}{(2\pi)^2} dx_1 dx_2$$

is invariant under the map $\mathbf{f}_{\alpha,p}$.
(ii) Find $\mathbf{f}_{\alpha,p}^{(n)}$, i.e. find the n-th iterate of $\mathbf{f}_{\alpha,p}$. Calculate

$$\mathbf{f}_{\alpha,p}(x_1, x_2) - \mathbf{f}_{\alpha,p}(y_1, y_2).$$

(iii) Discuss the map for $\alpha = 0$.
(iv) Let $L_2([0, 2\pi] \times [0, 2\pi])$ be the Hilbert space of the square integrable functions in the Lebesgue sense. Let $\langle\, , \rangle$ be the scalar product in this Hilbert space. We can define a unitary operator U associated with the invertible map $\mathbf{f}_{\alpha,p}$ by

$$Ug(x_1, x_2) := g(\mathbf{f}_{\alpha,p}(x_1, x_2)), \qquad g \in L_2([0, 2\pi] \times [0, 2\pi]).$$

Consider the *orthonormal basis*

$$\phi_{m,n}(x_1,x_2) = \frac{1}{2\pi}\exp(imx_1)\exp(inx_2), \qquad m,n \in \mathbb{Z}$$

in this Hilbert space $L_2([0,2\pi]\times[0,2\pi])$. Find the matrix representation of U.

Solution 22. (i) We have

$$d(x_1+\alpha) = dx_1, \qquad d(px_1+x_2) = pdx_1 + dx_2.$$

Since $dx_1 \wedge dx_1 = 0$ we have

$$d(x_1+\alpha)\wedge d(px_1+x_2) = dx_1 \wedge dx_2.$$

(ii) We compute the n-th iterate of $\mathbf{f}_{\alpha,p}^{(n)}$ and obtain

$$\mathbf{f}_{\alpha,p}^{(n)}(x_1,x_2) = [x_1+n\alpha, x_2+npx_1+n(n-1)p\alpha/2].$$

Thus for any couple of points (x_1,x_2) and (y_1,y_2) we have

$$\mathbf{f}_{\alpha,p}^{(n)}(x_1,x_2) - \mathbf{f}^{(n)}(y_1,y_2) = [x_1-y_1, x_2-y_2+np(x_1-y_1)] \ (\text{mod } 2\pi).$$

Therefore two arbitrary nearby points having distinct (x_1,y_1) coordinates will diverge according to a power law.
(iii) If $\alpha=0$ the dynamics consists in a rotation with a variable speed x_1. The system has a continuous spectrum. It admits a divergence of nearby trajectories, which is not of the exponential type.
(iv) We calculate the action of U on the orthonormal basis and obtain

$$\begin{aligned} U\phi_{n,m}(x_1,x_2) &= U\frac{1}{2\pi}e^{inx_1}e^{imx_2} = \frac{1}{2\pi}e^{in(x_1+\alpha)}e^{im(px_1+x_2)} \\ &= e^{in\alpha}\frac{1}{2\pi}e^{inx_1}e^{impx_1}e^{imx_2} = e^{in\alpha}\frac{1}{2\pi}e^{i(n+mp)x_1}e^{imx_2} \\ &= e^{in\alpha}\phi_{n+mp,m}(x_1,x_2) \end{aligned}$$

with $m,n\in\mathbb{Z}$.

Problem 23. Let $a>0$ and $a^2 = y_1y_0$. Apply the transformation

$$y_t = a\tan(x_t), \qquad t = 0,1,2,\ldots \tag{1}$$

to the second order nonlinear difference equation

$$y_{t+2}y_{t+1}y_t = a^2(y_{t+2}+y_{t+1}+y_t) \tag{2}$$

and thus find the solution of this second order difference equation. Note that

$$\tan(\alpha+\beta+\gamma) = \frac{\tan(\alpha)+\tan(\beta)+\tan(\gamma)-\tan(\alpha)\tan(\beta)\tan(\gamma)}{1-\tan(\alpha)\tan(\beta)-\tan(\alpha)\tan(\gamma)-\tan(\beta)\tan(\gamma)}.$$

Solution 23. We obtain

$$\tan(x_{t+2}+x_{t+1}+x_t) = 0$$

and hence the general solution of (2) is given by

$$y_t = a\tan\left(c_1\cos(2t\pi/3)+c_2\sin(2t\pi/3)+n\pi/2\right)$$

where c_1 and c_2 are arbitrary constants and n is an arbitrary integer.

Problem 24. Find the exact solution of the initial value problem for the system of difference equations

$$x_{1,t+1} = 2x_{1,t}^2 - 2x_{2,t}^2 - 1, \quad x_{2,t+1} = 4x_{1,t}x_{2,t}$$

where $t = 0, 1, \ldots$.

Solution 24. Using the addition theorems for sine, cosine, cosh and sinh we obtain

$$x_{1,t} = \cos(u_t)\cosh(v_t), \quad x_{2,t} = -\sin(u_t)\sinh(v_t)$$

where $u_t = 2^t u_0$ and $v_t = 2^t v_0$. The values u_0 and v_0 are determined by

$$x_{10} = \cos(u_0)\cosh(v_0), \quad x_{20} = -\sin(u_0)\sinh(v_0).$$

Given $x_{1,0}$ and $x_{2,0}$ this is a system of transcendental equations.

Problem 25. Find the exact solution of the initial value problem for the system of difference equations

$$x_{1,t+1} = 4x_{1,t}(1-x_{1,t}) + 4x_{2,t}^2, \quad x_{2,t+1} = 4x_{2,t}(1-2x_{1,t})$$

where $t = 0, 1, \ldots$.

Solution 25. Using the addition theorems for sine, cosine, cosh and sinh we obtain

$$x_{1,t} = \sin^2(u_t)\cosh^2(v_t) - \cos^2(u_t)\sinh^2(v_t)$$
$$x_{2,t} = 2\sin(u_t)\cos(u_t)\sinh(v_t)\cosh(v_t)$$

where $u_t = 2^t u_0$ and $v_t = 2^t v_0$. The values u_0 and v_0 are determined by

$$x_{1,0} = \sin^2(u_0)\cosh^2(v_0) - \cos^2(u_0)\sinh(v_0)$$
$$x_{2,0} = 2\sin(u_0)\cos(u_0)\sinh(v_0)\cosh(v_0).$$

Given $x_{1,0}$ and $x_{2,0}$ this is a system of transcendental equations.

Problem 26. Find the exact solution of the initial value problem for the system of difference equations

$$x_{1,t+1} = (2x_{1,t} - 2x_{2,t} - 1)(2x_{1,t} + 2x_{2,t} - 1), \quad x_{2,t+1} = 4x_{2,t}(2x_{1,t} - 1)$$

where $t = 0, 1, 2, \ldots$.

Solution 26. Using the addition theorems for sine, cosine, cosh and sinh we obtain

$$x_{1,t} = \cos^2(u_t)\cosh^2(v_t) - \sin^2(u_t)\sinh^2(v_t)$$
$$x_{2,t} = -2\sin(u_t)\cos(u_t)\sinh(v_t)\cosh(v_t)$$

where $u_t = 2^t u_0$ and $v_t = 2^t v_0$. The values u_0 and v_0 are determined by

$$x_{1,0} = \cos^2(u_0)\cosh^2(v_0) - \sin^2(u_0)\sinh(v_0)$$
$$x_{2,0} = -2\sin(u_0)\cos(u_0)\sinh(v_0)\cosh(v_0).$$

Given x_{10} and x_{20} this is a system of transcendental equations.

Problem 27. Let $n \geq 2$. An invertible integer matrix, $A \in GL_n(\mathbb{Z})$, generates a toral automorphism $f : \mathbb{T}^n \to \mathbb{T}^n$ via the formula

$$f \circ \pi = \pi \circ A, \qquad \pi : \mathbb{R}^n \to \mathbb{T}^n := \mathbb{R}^n/\mathbb{Z}^n.$$

The set of fixed points of f is given by

$$\text{Fix}(f) := \{\, x^* \in \mathbb{T}^n \;:\; f(x^*) = x^* \,\}.$$

Let $\sharp\text{Fix}(f)$ be the number of fixed points of f. Now we have: if

$$\det(I_n - A) \neq 0$$

then

$$\sharp\text{Fix}(f) = |\det(I_n - A)|.$$

Let $n = 2$ and

$$A = \begin{pmatrix} 2 & 1 \\ 1 & 1 \end{pmatrix}.$$

Show that $\det(I_2 - A) \neq 0$ and find $\sharp\text{Fix}(f)$.

Solution 27. We have $\det(I_2 - A) = -1$. Thus $|\det(I_2 - A)| = 1$.

Problem 28. Let $a, b > 0$. Find the condition on the bifurcation parameter b such that the map $\mathbf{f} : \mathbb{R}^2 \to \mathbb{R}^2$

$$f_1(x_1, x_2) = bx_2 + 1 - ax_1^2, \qquad f_2(x_1, x_2) = x_1$$

is area-preserving.

Solution 28. With $b = -1$ the map $\mathbf{f}$ is area-preserving, i.e. the determinant of

$$\begin{pmatrix} \partial f_1/\partial x_1 & \partial f_1/\partial x_2 \\ \partial f_2/\partial x_1 & \partial f_2/\partial x_2 \end{pmatrix} = \begin{pmatrix} -2ax_1 & b \\ 1 & 0 \end{pmatrix}$$

is equal to 1.

Problem 29. Let $r > 0$. Find the fixed points and their stability of the two-dimensional map

$$\begin{aligned} x_{1,t+1} &= r(3x_{2,t} + 1)x_{1,t}(1 - x_{1,t}) \\ x_{2,t+1} &= r(3x_{1,t} + 1)x_{2,t}(1 - x_{2,t}). \end{aligned}$$

Study the stability of the fixed point $(0, 0)$.

Solution 29. The variational equation takes the form

$$\begin{aligned} y_{1,t+1} &= r(3x_{2,t} + 1)(1 - 2x_{1,t})y_{1,t} + 3rx_{1,t}(1 - x_{1,t})y_{2,t} \\ y_{2,t+1} &= 3rx_{2,t}(1 - x_{2,t})y_{1,t} + r(3x_{1,t} + 1)(1 - 2x_{2,t})y_{2,t}. \end{aligned}$$

For the fixed point $(0, 0)$ we have

$$\frac{\partial f_1(0,0)}{\partial x_1} = r, \quad \frac{\partial f_1(0,0)}{\partial x_2} = 0, \quad \frac{\partial f_2(0,0)}{\partial x_1} = 0, \quad \frac{\partial f_2(0,0)}{\partial x_2} = r$$

and therefore the variational equation reduces to

$$\begin{pmatrix} y_{1,t+1} \\ y_{2,t+1} \end{pmatrix} = \begin{pmatrix} r & 0 \\ 0 & r \end{pmatrix} \begin{pmatrix} y_{1,t} \\ y_{2,t} \end{pmatrix}.$$

The eigenvalues of the matrix are r (twice). Thus the fixed point is stable for $r < 1$ and unstable for $r > 1$.

Problem 30. (i) Consider the differential two-form in $\mathbb{R}^2$

$$\omega = dx_1 \wedge dx_2.$$

Find all smooth maps $\mathbf{f} : \mathbb{R}^2 \to \mathbb{R}^2$ which leaves ω invariant.
(ii) Consider the differential one-form in $\mathbb{R}^2$

$$\alpha = x_1 dx_2 - x_2 dx_1.$$

Find all smooth maps $\mathbf{f} : \mathbb{R}^2 \to \mathbb{R}^2$ which leaves α invariant. Note that $d\alpha = 2dx_1 \wedge dx_2$.

Solution 30. (i) Since

$$df_1 = \frac{\partial f_1}{\partial x_1} dx_1 + \frac{\partial f_1}{\partial x_2} dx_2, \quad df_2 = \frac{\partial f_2}{\partial x_1} dx_1 + \frac{\partial f_2}{\partial x_2} dx_2$$

we find

$$\mathbf{f}^*(\omega) = df_1 \wedge df_2 = \left(\frac{\partial f_1}{\partial x_1}\frac{\partial f_2}{\partial x_2} - \frac{\partial f_1}{\partial x_2}\frac{\partial f_2}{\partial x_1} \right) dx_1 \wedge dx_2.$$

This provides the condition

$$\left(\frac{\partial f_1}{\partial x_1}\frac{\partial f_2}{\partial x_2} - \frac{\partial f_1}{\partial x_2}\frac{\partial f_2}{\partial x_1} \right) = 1.$$

(ii) Since

$$f_1 \left(\frac{\partial f_2}{\partial x_1} dx_1 + \frac{\partial f_2}{\partial x_1} dx_2 \right) - f_2 \left(\frac{\partial f_1}{\partial x_1} dx_1 + \frac{\partial f_1}{\partial x_2} dx_2 \right) = x_1 dx_2 - x_2 dx_1$$

we obtain the two conditions

$$f_1 \frac{\partial f_2}{\partial x_1} - f_2 \frac{\partial f_1}{\partial x_1} = -x_2, \qquad f_1 \frac{\partial f_2}{\partial x_2} - f_2 \frac{\partial f_1}{\partial x_2} = x_1.$$

A consequence of the two conditions is that

$$f_1 \frac{\partial^2 f_2}{\partial x_1 \partial x_2} = f_2 \frac{\partial^2 f_1}{\partial x_1 \partial x_2}.$$

Problem 31. Let $r > 0$ be the bifurcation parameter. The *whisker map* is given by

$$f_1(x, \theta) = x + 4r\sin(\theta), \qquad f_2(x, \theta) = \theta - \ln(|1 + x + 4r\sin(\theta)|) \mod 2\pi.$$

Find the fixed points and study their stability.

Solution 31. We have to consider the system of equations

$$x + 4r\sin(\theta) = x, \qquad \theta = \theta - \ln(|1 + x + 4r\sin(\theta)|).$$

to find the fixed points. From the first equation we find $\theta = n\pi$ with $n \in \mathbb{Z}$. The second equation provides the restriction that $\theta = 0, \pi, -\pi$. With $\ln(|1|) = 0$ we obtain the fixed points

$$(0,0), \quad (0,\pi), \quad (0,-\pi)$$

and

$$(-2,0), \quad (-2,\pi), \quad (-2,-\pi).$$

Problem 32. Let $r > 0$. Consider the autonomous system of differential equations

$$\frac{du_1}{dt} = u_2, \qquad \frac{du_2}{dt} = ru_1(u_1 - 1)$$

with the fixed points $(0,0)$ and $(1,0)$ as solutions of

$$u_2^* = 0, \qquad ru_1^*(u_1^* - 1) = 0.$$

The corresponding smooth *vector field* V in $\mathbb{R}^2$ for the system of differential equation is

$$V = u_2\frac{\partial}{\partial u_1} + ru_1(u_1 - 1)\frac{\partial}{\partial u_2}.$$

We have $Vu_1 = u_2$, $Vu_2 = ru_1(u_1 - 1)$. We have for the *Lie series*

$$e^{tV}u_1 = u_1 + tu_2 + \cdots$$

$$e^{tV}u_2 = u_2 + tru_1(u_1 - 1) + \cdots.$$

Motivated by the Lie series expansion for the solution of the system of differential equations and truncation with $t = 1$ we replace the system of differential equations by the two-dimensional map

$$f_1(x_1, x_2) = x_1 + x_2, \qquad f_2(x_1, x_2) = x_2 + rx_1(x_1 - 1).$$

Find the fixed points of this map.

Solution 32. There are two fixed points at $(0,0)$ and $(1,0)$. These fixed points are the same fixed points of system of differential equation.

Problem 33. Let H be the Heaviside step function, i.e. $H(x) = 0$ for $x < 0$ and $H(x) = 1$ for $x \geq 0$. Consider the two-dimensional map (butcher's map)

$$\begin{pmatrix} x_{1,t+1} \\ x_{2,t+1} \end{pmatrix} = \begin{pmatrix} 1/2 & 0 \\ 0 & 2 \end{pmatrix} \begin{pmatrix} x_{1,t} \\ x_{2,t} \end{pmatrix} + \begin{pmatrix} (1/2 + \epsilon)H(2x_{2,t} - 1) \\ 0 \end{pmatrix} \mod 1.$$

Discuss the behaviour of the map.

Solution 33. The butcher's map stretches in the x_2-direction but squeezes in the x_1-direction. Thus the area is locally preserved. The butcher's map is not invertible. After an iteration a gap and an overlap is found. The butcher's map admits an attractor.

Problem 34. Let $k \geq 0$ be the bifurcation parameter. Discuss the behaviour of the two-dimensional map

$$\theta_{t+1} = \theta_t + r_{t+1} \mod 1$$
$$r_{t+1} = r_t - \frac{k}{2\pi} \sin(2\pi\theta_t)$$

in dependence of the bifurcation parameter k, where $t = 0, 1, \ldots$.

Solution 34. If $k = 0$, i.e. $\theta_{t+1} = \theta_t + r_{t+1}$, $r_{t+1} = r_t$ the phase space is filled up by tori connecting $\theta = 0$ to $\theta = 1$. When k is increased these tori break down more and more and chaotic regions appear. If $k > k_c = 0.971635406$ all the tori connecting $\theta = 0$ to $\theta = 1$ vanish. A torus which has the golden-mean rotation number ρ_G survives at the bifurcation point $k = k_c$. It is called the critical KAM torus.

Problem 35. Let $r > 0$ be the bifurcation parameter. Study the map $\mathbf{f} : \mathbb{R}^2 \to \mathbb{R}^2$

$$f_1(x_1, x_2) = -x_2 + x_1(rx_1^2 + 2 - r), \quad f_2(x_1, x_2) = x_1$$

or written as difference equations

$$x_{1,t+1} = -x_{2,t} + x_{1,t}(rx_{1,t}^2 + 2 - r), \quad x_{2,t+1} = x_{1,t}.$$

Solution 35. The map is area preserving. The fixed point is $(0, 0)$. The fixed point bifurcates into a 2-cycle orbit.

Problem 36. Let $a > 0$, $r > 0$. Consider the map (*delayed logistic map*)

$$f_1(x_1, x_2) = ax_1 + rx_2(1 - x_2), \quad f_2(x_1, x_2) = x_1.$$

(i) Find $\mathbf{f}^*(dx_1 \wedge dx_2)$.
(ii) Show that $(0, 0)$ is a fixed point.

Solution 36. (i) Since

$$df_1 = adx_1 + rdx_2 - 2rx_2dx_2, \quad df_2 = dx_1$$

and $dx_1 \wedge dx_1 = 0$, $dx_2 \wedge dx_2 = 0$ we have

$$\mathbf{f}^*(dx_1 \wedge dx_2) = r dx_2 \wedge dx_1 - 2rx_2 dx_2 \wedge dx_1 = r(2x_2 - 1) dx_1 \wedge dx_2.$$

(ii) From

$$ax_1 + rx_2(1 - x_2) = 0, \qquad x_1 = x_2$$

we obtain $ax_1 + rx_1(1 - x_1) = 0$ which implies that $(x_1, x_2) = (0, 0)$ is a fixed point and $(x_1, x_2) = ((a + r)/2, (a + r)/2)$ is a fixed point.

Let $a = 0.5$ and $r \in [1, 4]$. Show that one has a transition

fixed point, Hopf bifurcation, torus, locking, chaos, hyperchaos

Problem 37. (i) Study the two-dimensional map ($t = 0, 1, 2, \ldots$)

$$x_{1,t+1} = x_{1,t} + \sin(\pi x_{1,t}) \sin(\pi x_{2,t}), \quad x_{2,t+1} = x_{2,t} + \sin(\pi x_{1,t}) \sin(\pi x_{2,t})$$

with the initial values x_0 and y_0.
(ii) Study the two-dimensional map ($t = 0, 1, 2, \ldots$)

$$x_{1,t+1} = x_{1,t} + \sin(\pi x_{1,t}) \sin(\pi x_{2,t}), \quad x_{2,t+1} = x_{2,t} - \sin(\pi x_{1,t}) \sin(\pi x_{2,t})$$

with the initial values $x_{1,0}$ and $x_{2,0}$.

Solution 37. (i) We have $x_{1,t+1} - x_{2,t+1} = x_{1,t} - x_{2,t}$.
(ii) We have $x_{1,t+1} + x_{2,t+1} = x_{1,t} + x_{2,t}$.

Problem 38. Let $a > 0$ and $\sigma \geq 0$. Consider the potential

$$U(q) = (q - a)^2 (q + a)^2 \equiv q^4 - 2a^2 q^2 + a^4.$$

Then

$$\frac{dU(q)}{dq} = 4q^3 - 4a^2 q.$$

Consider the two-dimensional map

$$q_{t+1} = q_t + \sigma p_{t+1}$$
$$p_{t+1} = p_t - \sigma \frac{dU(q = q_t)}{dq} = p_t - 4\sigma q_t (q_t^2 - a^2).$$

(i) Find the Jacobian matrix

$$J = \begin{pmatrix} \partial q_{t+1}/\partial q_t & \partial q_{t+1}/\partial p_t \\ \partial p_{t+1}/\partial q_t & \partial p_{t+1}/\partial p_t \end{pmatrix}.$$

Then find $\det(J)$.
(ii) Let

$$J = \begin{pmatrix} 0 & 1 \\ -1 & 0 \end{pmatrix}.$$

Find $M^T JM$, where ^{T} denotes the transpose.
(iii) Find the fixed points of the map and study their stability.

Solution 38. (i) We obtain the Jacobian matrix

$$J = \begin{pmatrix} \partial q_{t+1}/\partial q_t & \partial q_{t+1}/\partial p_t \\ \partial p_{t+1}/\partial q_t & \partial p_{t+1}/\partial p_t \end{pmatrix} = \begin{pmatrix} 1 - 4\sigma^2(3q_t^2 - a^2) & \sigma \\ -4\sigma(3q_t^2 - a^2) & 1 \end{pmatrix}.$$

It follows that $\det(J) = 1$, i.e. the map is area-preserving.
(ii) We obtain $M^T JM = J$, i.e. the map satisfies the simplecticity condition.
(iii) From

$$\begin{aligned} q^* &= q^* + \sigma(p^* - 4\sigma q^*(q^* + a)(q^* - a) \\ p^* &= p^* - 4\sigma q^*(q^* + a)(q^* - a) \end{aligned}$$

we obtain

$$0 = \sigma(p^* - 4\sigma q^*(q^* + a)(q^* - a), \quad 0 = -4\sigma q^*(q^* + a)(q^* - a)$$

and thus we find three fixed points. The fixed point at $(q^*, p^*) = (0, 0)$ is hyperbolic and the other two fixed points at $(a, 0)$ and $(-a, 0)$ are elliptic.

Problem 39. Study the initial value problem of the difference equation

$$\det \begin{pmatrix} x_{t+2} & x_{t+1} \\ x_{t+1} & x_t \end{pmatrix} = 0$$

where $t = 0, 1, 2, \ldots$ and $x_0 = 1$, $x_1 = 1/2$.

Solution 39. We have $x_{t+2}x_t - x_{t+1}x_{t+1} = 0$. Thus

$$x_0 = 1, \; x_1 = 1/2, \; x_2 = 1/2^2, \ldots, x_t = 1/2^t.$$

The sequence tends to 0.

Problem 40. Let $\{\, , \,\}$ be the *Poisson bracket*. It is defined by

$$\{Q(p, q), P(p, q)\} := \frac{\partial Q}{\partial q}\frac{\partial P}{\partial p} - \frac{\partial Q}{\partial p}\frac{\partial P}{\partial q}.$$

Consider the transformation

$$Q(p,q) = f(q+yp) + aq + bq$$
$$P(p,q) = xf(q+yp) + cq + dp$$

with $ad - bc = 1$, f is a differentiable function of $q + yp$ and x, y are real numbers. Show that the transformation is canonical if and only if x, y are connected by

$$y = \frac{bx - d}{ax - c}.$$

Solution 40. We have

$$\{Q(p,q), P(p,q)\} = (c - ax)\{f,q\} + (d - bx)\{f,p\} + (ad - bc)\{q,p\}$$
$$= ((ax - c)\frac{\partial f}{\partial p} - (bx - d)\frac{\partial f}{\partial q} + 1)\{q,p\} = \{q,p\}$$

iff $y(ax - c) = bx - d$.

Problem 41. Let (open disc)

$$D^2 := \{\, (x_1, x_2) \in \mathbb{R}^2 \;:\; x_1^2 + x_2^2 < 1 \,\}.$$

Show that D^2 is *homeomorphic* to $\mathbb{R}^2$.

Solution 41. A homeomorphism $\mathbf{f} : D^2 \to \mathbb{R}^2$ is

$$\mathbf{f}(x_1, x_2) = \left(\frac{x_1}{\sqrt{1 - x_1^2 - x_2^2}}, \frac{x_2}{\sqrt{1 - x_1^2 - x_2^2}} \right).$$

Then the inverse $\mathbf{f}^{-1} : \mathbb{R}^2 \to D^2$ is given by

$$\mathbf{f}^{-1}(x_1, x_2) = \left(\frac{x_1}{\sqrt{1 + x_1^2 + x_2^2}}, \frac{x_2}{\sqrt{1 + x_1^2 + x_2^2}} \right).$$

Problem 42. Let $r \geq 0$. Consider the two-dimensional invertible map

$$x_{1,t+1} = \frac{3}{2}x_1 + 2x_1^2 - \frac{1}{2}x_{2,t} + r, \quad x_{2,t+1} = x_{1,t}, \qquad t = 0, 1, 2, \ldots$$

Find the fixed points. Discuss.

Solution 42. Solving

$$\frac{3}{2}x_1^* + 2(x_1^*)^2 - \frac{1}{2}x_2^* + r = x_1^*, \quad x_1^* = x_2^*$$

shows that the map has two complex fixed points if $r > 0$ at a distance proportional to $\sqrt{r}$ from the real plane. We find

$$(x_1^* = -\sqrt{r}i/\sqrt{2}, x_2^* = -\sqrt{r}i/\sqrt{2}), \qquad (x_1^* = \sqrt{r}i/\sqrt{2}, x_2^* = \sqrt{r}i/\sqrt{2}).$$

No real fixed points exist for $r > 0$. For $r = 0$ we obtain the fixed point $(0, 0)$. The map admits a saddle-node bifurcation if the bifurcation parameter r approaches 0 from above.

Problem 43. Consider the smooth *vector field* in $\mathbb{R}^2$

$$V = (x_1 - x_2)^2 \frac{\partial}{\partial x_1} + (x_1 - x_2)^2 \frac{\partial}{\partial x_2} \equiv (x_1 - x_2)^2 \left(\frac{\partial}{\partial x_1} + \frac{\partial}{\partial x_2} \right).$$

Find the map $\mathbf{f} : \mathbb{R}^2 \to \mathbb{R}^2$ generated by the *Lie series*

$$f_1(x_1, x_2) = x_1' = e^V x_1, \qquad f_2(x_1, x_2) = x_2' = e^V x_2.$$

Find a fixed point of this map.

Solution 43. Since $Vx_1 = (x_1 - x_2)$, $Vx_2 = (x_1 - x_2)^2$ and

$$V(Vx_1) = 0, \qquad V(Vx_2) = 0$$

we obtain

$$f_1(x_1, x_2) = x_1 + (x_1 - x_2)^2, \qquad f_2(x_1, x_2) = x_2 + (x_1 - x_2)^2.$$

A fixed point is given by $(x_1^*, x_2^*) = (0, 0)$. This is also a fixed point of the system of differential equations

$$\frac{dx_1}{dt} = (x_1 - x_2)^2, \quad \frac{dx_2}{dt} = (x_1 - x_2)^2.$$

Problem 44. Let $\alpha \in \mathbb{R}$. Consider the two-dimensional map $\mathbf{f} : \mathbb{R}^2 \to \mathbb{R}^2$

$$f_1(x_1, x_2) = (x_1 + x_2^3) \cos(\alpha) - x_2 \sin(\alpha)$$
$$f_2(x_1, x_2) = (x_1 + x_2^3) \sin(\alpha) + x_2 \cos(\alpha).$$

Find the inverse of the map if it exists. Is there a fixed point for all α.

Solution 44. Note that

$$\det \begin{pmatrix} \partial f_1/\partial x_1 & \partial f_1/\partial x_2 \\ \partial f_2/\partial x_1 & \partial f_2/\partial x_2 \end{pmatrix} = 1.$$

Thus the inverse of $\mathbf{f}$ exists and is given by

$$f_1^{-1}(x_1, x_2) = x_1 \cos(\alpha) + x_2 \sin(\alpha) + (x_1 \sin(\alpha) - x_2 \cos(\alpha))^3$$
$$f_2^{-1}(x_1, x_2) = -x_1 \sin(\alpha) + x_2 \cos(\alpha).$$

Yes obviously $(x_1^*, x_2^*) = (0, 0)$ is such a fixed point.

Problem 45. Let $g : \mathbb{R} \to \mathbb{R}$ be an analytic map. Consider the map $\mathbf{f} : \mathbb{R}^2 \to \mathbb{R}^2$

$$f_1(x_1, x_2) = g(x_1) + r(g(x_2) - g(x_1)), \quad f_2(x_1, x_2) = g(x_2) + r(g(x_1) - g(x_2))$$

or written as a system of difference equations

$$x_{1,t+1} = g(x_{1,t}) + r(g(x_{2,t}) - g(x_{1,t})), \quad x_{2,t+1} = g(x_{2,t}) + r(g(x_{1,t}) - g(x_{2,t})).$$

Note that

$$x_{1,t+1} + x_{2,t+1} = g(x_{1,t}) + g(x_{2,t})$$
$$x_{1,t+1} - x_{2,t+1} = g(x_{1,t})(1 - 2r) - g(x_{2,t})(1 - 2r).$$

Let

$$g(x) = 1 - 2x^2.$$

We define $S_t := \frac{1}{2}(x_{1,t} + x_{2,y}$ and $D_t := \frac{1}{2}(x_{1,t} - x_{2,t})$. Find the time evolution of S_t and D_t.

Solution 45. We have

$$S_{t+1} = \frac{1}{2}(g(x_{1,t}) + g(x_{2,t}))$$
$$D_{t+1} = \frac{1}{2}(g(x_{1,t}) - g(x_{2,t}) + rg(x_{2,t}) - rg(x_{1,t}).$$

With $g(x) = 1 - 2x^2$ we obtain

$$S_{t+1} = 1 - x_{1,t}^2 - x_{2,t}^2, \quad D_{t+1} = (-1 + 2r)(x_{1,t}^2 - x_{2,t}^2).$$

Since

$$S_t^2 = \frac{1}{4}(x_{1,t}^2 + x_{2,t}^2 + 2x_{1,t}x_{2,t}), \quad D_t^2 = \frac{1}{4}(x_{1,t}^2 + x_{2,t}^2 - 2x_{1,t}x_{2,t})$$

and

$$D_t S_t = \frac{1}{4}(x_{1,t}^2 - x_{2,t}^2)$$

we finally arrive at

$$S_{t+1} = 1 - 2(S_t^2 + D_t^2), \qquad D_{t+1} = 4(-1 + 2r)S_t D_t.$$

Problem 46. Consider the manifold $M = \mathbb{R}^2$ and the differential two-form $\omega = dx_1 \wedge dx_2$ and the differential one form $\alpha = x_1 dx_2 - x_2 dx_1$. Express these differential forms in *polar coordinates* $x_1(r,\theta) = r\cos(\theta)$, $x_2(r,\theta) = r\sin(\theta)$.

Solution 46. Since

$$dx_1(r,\theta) = \cos(\theta)dr - r\sin(\theta)d\theta$$
$$dx_2(r,\theta) = \sin(\theta)dr + r\cos(\theta)d\theta$$

we obtain

$$dx_1 \wedge dx_2 = r\cos^2(\theta)dr \wedge d\theta - r\sin^2(\theta)d\theta \wedge dr = rdr \wedge d\theta.$$

For the differential one form α we find

$$\begin{aligned}\alpha &= r\cos(\theta)d(r\sin(\theta)) - r\sin(\theta)d(r\cos(\theta))\\ &= r\cos(\theta)(\sin(\theta)dr + r\cos(\theta)d\theta) - r\sin(\theta)(\cos(\theta)dr - r\sin(\theta)d\theta)\\ &= r^2 d\theta.\end{aligned}$$

Problem 47. Consider the manifold $M = \mathbb{R}^2$ and the differential two-form $\omega = dx_1 \wedge dx_2$. Let $\mathbf{f} : \mathbb{R}^2 \to \mathbb{R}^2$ be a smooth function and V be a smooth vector field

$$V = V_1(x_1,x_2)\frac{\partial}{\partial x_1} + V_2(x_1,x_2)\frac{\partial}{\partial x_2}$$

in $\mathbb{R}^2$. Find the conditions on f_1, f_2, V_1, V_2 such that

$$L_V\omega = \mathbf{f}^*(\omega)$$

where $L_V(.)$ denotes the *Lie derivative.*

Solution 47. Since

$$df_1 = \frac{\partial f_1}{\partial x_1}dx_1 + \frac{\partial f_1}{\partial x_2}dx_2, \quad df_2 = \frac{\partial f_2}{\partial x_1}dx_1 + \frac{\partial f_2}{\partial x_2}dx_2$$

we obtain utilizing $dx_1 \wedge dx_1 = dx_2 \wedge dx_2 = 0$ and $dx_1 \wedge dx_2 = -dx_2 \wedge dx_1$

$$\mathbf{f}^*(\omega) = df_1 \wedge df_2 = \left(\frac{\partial f_1}{\partial x_1}\frac{\partial f_2}{\partial x_2} - \frac{\partial f_1}{\partial x_2}\frac{\partial f_2}{\partial x_1}\right)dx_1 \wedge dx_2.$$

Note that the Lie derivative L_V and the *exterior derivative* d commute. Thus for the Lie derivative we find

$$L_V(dx_1 \wedge dx_2) = d(L_V x_1)\wedge dx_2 + dx_1 \wedge d(L_V x_2) = \left(\frac{\partial V_1}{\partial x_1} + \frac{\partial V_2}{\partial x_2}\right)dx_1 \wedge dx_2.$$

Hence the condition is

$$\left(\frac{\partial V_1}{\partial x_1} + \frac{\partial V_2}{\partial x_2}\right) = \left(\frac{\partial f_1}{\partial x_1}\frac{\partial f_2}{\partial x_2} - \frac{\partial f_1}{\partial x_2}\frac{\partial f_2}{\partial x_1}\right).$$

Problem 48. Let $r_1 > 0$, $r_2 > 0$. Consider the analytic function $\mathbf{f} : \mathbb{R}^2 \to \mathbb{R}^2$

$$f_1(x_1, x_2) = r_1 x_1 (1 - x_1 - x_2), \qquad f_2(x_1, x_2) = r_2 x_1 x_2.$$

Find df_1, df_2 and $df_1 \wedge df_2$.

Solution 48. We have $\mathbf{f} : \mathbb{R}^2 \to \mathbb{R}^2$

$$df_1(x_1, x_2) = r_1(1 - 2x_1 - x_2)dx_1 - r_1 x_1 dx_2$$
$$df_2(x_1, x_2) = r_2 dx_1 + r_2 x_1 dx_2.$$

It follows that

$$df_1 \wedge df_2 = r_1 r_2 x_1 (1 - 2x_1) dx_1 \wedge dx_2.$$

It follows that the map is not invertible.

Problem 49. Consider the analytic function $\mathbf{f} : \mathbb{R}^2 \to \mathbb{R}^2$

$$f_1(x_1, x_2) = x_1 + x_2, \qquad f_2(x_1, x_2) = x_1 x_2.$$

(i) Find (function composition)

$$g_1(x_1, x_2) = f_1(f_1(x_1, x_2), f_2(x_1, x_2))$$
$$g_2(x_1, x_2) = f_2(f_1(x_1, x_2), f_2(x_1, x_2)).$$

Apply computer algebra.
(ii) Find the commutator

$$f_1(f_1(x_1, x_2), f_2(x_1, x_2)) - f_2(f_1(x_1, x_2), f_2(x_1, x_2)).$$

Apply computer algebra.

Solution 49. (i) A SymbolicC++ program is

```
// composition.cpp

#include <iostream>
#include "symbolicc++.h"
```

```
using namespace std;

int main(void)
{
  Symbolic f1("f1"), f2("f2"), x1("x1"), x2("x2");
  Symbolic t("t");
  f1 = x1 + x2;
  f2 = x1*x2;
  cout << f1 << endl;
  cout << f2 << endl;
  Symbolic g1("g1"), g2("g2");
  g1 = f1[x2==t,x1==f1,t==f2,t==x2];
  cout << g1 << endl;
  g2 = f2[x2==t,x1==f1,t==f2,t==x2];
  cout << g2 << endl;
  return 0;
}
```

The output is

```
x1+x2
x1*x2
x1+x2+x1*x2
x1^(2)*x2+x2^(2)*x1
```

(ii) A SymbolicC++ program is

```
// fcommutator.cpp

#include <iostream>
#include "symbolicc++.h"
using namespace std;

Symbolic f1(const Symbolic& x1,const Symbolic& x2)
         { return x1 + x2; }

Symbolic f2(const Symbolic& x1,const Symbolic& x2)
         { return x1*x2; }

Symbolic C(Symbolic (*f1)(const Symbolic&,const Symbolic&),
           Symbolic (*f2)(const Symbolic&,const Symbolic&),
           const Symbolic& x1,const Symbolic& x2)
{ return (f1(f1(x1,x2),f2(x1,x2)) - f2(f1(x1,x2),f2(x1,x2))); }

int main(void)
{
  Symbolic x1("x1"), x2("x2");
  cout << C(f1,f2,x1,x2) << endl;
  return 0;
}
```

The output is

```
x1+x2+x1*x2-x1^(2)*x2-x2^(2)*x1
```

Problem 50. Let $r \geq 0$. Study Hopf bifurcation for the two-dimensional map

$$f_1(x_1, x_2) = rx_1(3x_2 + 1)(1 - x_1), \qquad f_2(x_1, x_2) = rx_2(3x_1 + 1)(1 - x_2)$$

or written as system of difference equations

$$x_{1,t+1} = r(3x_{2,t} + 1)x_{1,t}(1 - x_{1,t}), \qquad x_{2,t+1} = r(3x_{1,t} + 1)x_{2,t}(1 - x_{2,t}).$$

First find the fixed points.

Solution 50. The fixed points are given as the solution of the equations

$$x_1^* = r(3x_2^* + 1)x_1^*(1 - x_1^*), \qquad x_2^* = r(3x_1^* + 1)x_2^*(1 - x_1^*).$$

Obviously $\mathbf{x}_0^* = (0, 0)$ is a fixed point for all r. For $r \neq 0$ we find the fixed points

$$\mathbf{x}_1^* = ((r-1)/r, 0), \quad \mathbf{x}_2^* = (0, (r-1)/r).$$

For $(r \geq 3/4)$ we find two more fixed points on the diagonal

$$\mathbf{x}_3^* = (1/3 - \sqrt{4 - 3/r}/3, 1/3 - \sqrt{4 - 3/r})$$
$$\mathbf{x}_4^* = (1/3 + \sqrt{4 - 3/r}/3, 1/3 + \sqrt{4 - 3/r}).$$

To study Hopf bifurcation consider these two fixed points $\mathbf{x}_3^*$ and $\mathbf{x}_4^*$ on the diagonal. These fixed points exists only for $r \geq 3/4$. For $r > 3/4$ a stable period-2 orbit exists. This period-2 orbit looses stability via a Hopf bifurcation which occurs at $r = r_0$, where $r_0 = 0.957$ and gives rise to a stable limit cycle for $r \in [r_0, r_0 + \delta)$ for some $\delta > 0$.

Problem 51. Let n be a natural number with $n \geq 2$. We set

$$x_0 = 0, \quad y_0 = x_1 = 1, \quad y_1 = n$$

and

$$x_{t+2} = \left\lfloor \frac{y_t + n}{y_{t+1}} \right\rfloor - x_t, \quad y_{t+2} = \left\lfloor \frac{y_t + n}{y_{t+1}} \right\rfloor - y_t$$

where $\lfloor a \rfloor$ denotes the greatest integer not greater than a. The ratio x_t/y_t is called the *Farey fraction* and

$$\frac{x_0}{y_0}, \ \frac{x_1}{y_1}, \ \frac{x_2}{y_2}, \ \ldots$$

is called the Farey sequence. Let $n = 5$. Find the sequence.

Solution 51. We obtain in the interval

$$\frac{0}{1}, \frac{1}{5}, \frac{1}{4}, \frac{1}{3}, \frac{2}{5}, \frac{1}{2}, \frac{3}{5}, \frac{2}{3}, \frac{3}{4}, \frac{4}{5}, \frac{1}{1}.$$

Problem 52. Let n be a natural number. The recursive relation used to determine the *Farey fraction* x_k/y_k is given by

$$x_{k+2} = \left\lfloor \frac{y_k + n}{y_{k+1}} \right\rfloor x_{k+1} - x_k, \quad y_{k+2} = \left\lfloor \frac{y_k + n}{y_{k+1}} \right\rfloor y_{k+1} - y_k$$

where the initial conditions are $x_0 = 0$, $y_0 = x_1 = 1$ and $y_1 = n$. The sequence of x_k/y_k is called the *Farey sequence.* The floor of a denoted by $\lfloor a \rfloor$ is the greatest integer which is not greater than a, for example $\lfloor 5.3 \rfloor = 5$. Write a C++ program to determine the Farey sequence for given n. Determine the first 11 elements of the sequence for $n = 5$.

Solution 52.

```
// Farey.cpp

#include <iostream>
#include <cmath>
using namespace std;

int main(void)
{
  int n = 5;
  int xk = 0, yk = 1;
  int xk1 = 1, yk1 = n;
  int xk2, yk2;
  int m = 11;
  for(;m>0;m--)
  {
  xk2 = (int) floor(double(yk+n)/yk1)*xk1-xk;
  yk2 = (int) floor(double(yk+n)/yk1)*yk1-yk;
  cout << xk << "/" << yk << endl;
  xk = xk1; yk = yk1; xk1 = xk2; yk1 = yk2;
  }
  return 0;
}
```

The output is

```
0/1
```

```
1/5
1/4
1/3
2/5
1/2
3/5
2/3
3/4
4/5
1/1
```

2.2.2 Supplementary Problems

Problem 1. (i) Study the coupled map

$$x_{1,t+1} = 2x_{2,t} \mod 1, \qquad x_{2,t+1} = 2x_{1,t}x_{2,t} \mod 1$$

where $t = 0, 1, 2, \ldots$ and $x_{1,0}, x_{2,0} \in [0, 1)$. Consider the cases that $x_{1,0}, x_{2,0}$ rational and irrational numbers in the interval $[0, 1)$.
(ii) Study the coupled map

$$x_{1,t+1} = 2x_{1,t}x_{2,t} \mod 1, \qquad x_{2,t+1} = 2x_{1,t}x_{2,t} \mod 1$$

where $t = 0, 1, 2, \ldots$ and $x_{1,0}, x_{2,0} \in [0, 1)$. Consider the cases that $x_{1,0}, x_{2,0}$ rational and irrational numbers in the interval $[0, 1)$.
(iii) Study the coupled map

$$x_{1,t+1} = 2x_{2,t} \mod 1, \qquad x_{2,t+1} = 2x_{1,t} \mod 1$$

where $t = 0, 1, 2, \ldots$ and $x_{1,0}, x_{2,0} \in [0, 1)$. Find the orbit for the initial values $x_{1,0} = 1/4$, $x_{2,0} = 3/4$. Study the cases that $x_{1,0}, x_{2,0}$ rational and irrational numbers in the interval $[0, 1)$.

Problem 2. Let $a > 0$, $b > 0$. Study the *Kaplan-Yorke map* defined by

$$x_{1,t+1} = ax_{1,t} \mod 1, \quad x_{2,t+1} = bx_{2,t} + \cos(2\pi x_{1,t}); \quad t = 0, 1, 2, \ldots$$

Problem 3. Let $a > 0$, $b > 0$. Consider the two-dimensional map

$$x_{1,t+1} = x_{1,t} + a(\exp(x_{2,t} - x_{1,t}) - 1), \quad x_{2,t+1} = x_{2,t} + b(x_{1,t} - x_{2,t})$$

Show that the fixed point are given by the manifold $x_1^* = x_2^*$. Show that the local stability condition is given by $a + b < 2$.

Problem 4. Let $\alpha \in \mathbb{R}$. Study the quadratic *Cremona map*

$$f_1(x_1, x_2) = x_1 \cos(\alpha) + x_2 \sin(\alpha) + x_2^2 \cos(\alpha)$$
$$f_2(x_1, x_2) = -x_1 \sin(\alpha) + x_2 \cos(\alpha) - x_2^2 \sin(\alpha)$$

which admits the fixed point $(0,0)$ for all α. Is the fixed point stable?

Problem 5. The *Chebyshev polynomials* $T_k(x)$ may be defined by the expression

$$T_k(x) := \cos(k \cos^{-1}(x)) \tag{1}$$

where $k \in \mathbb{N}_0$ and $|x| \le 1$.
(i) Show that

$$T_{n+m}(x) + T_{n-m}(x) = 2T_n(x)T_m(x), \qquad n \le m.$$

(ii) Show that

$$T_{k+1}(x) - xT_k(x) + \frac{1}{4}T_{k-1}(x) = 0.$$

Problem 6. (i) Study the second order difference equation

$$x_{t+1} - 2x_t + x_{t-1} = x_t(1 - x_t), \qquad t = 1, 2, \ldots$$

with given $x_0 > 0$ and $x_1 > 0$ (initial value problem).
(ii) Let $r > 0$. Study the second order difference equation

$$x_{t+1} + x_{t-1} = \frac{rx_t}{1 + x_t^2}, \qquad t = 1, 2, \ldots$$

with given $x_1, x_0 > 0$ (initial value problem).

Problem 7. Let $r > 0$, $a > 0$, $b > 0$, $c > 0$. Consider the system of difference equations

$$x_{1,t+1} = x_{1,t} + rx_{1,t}(a - b(2x_{1,t} + x_{2,t}) - 2cx_{1,t})$$
$$x_{2,t+1} = x_{2,t} + rx_{2,t}(a - b(x_{1,t} + 2x_{2,t}) - 2cx_{2,t}).$$

Show that this system admits the four fixed points

$$(0,0), \;\; (a/(2(b+c)), 0), \;\; (0, a/(2(b+c)), \;\; (a/(3b+2c)), a/(3b+2c))).$$

These four fixed points can be found with the Maxima program

```
ex1: k*x1*(a-b*(2*x1+x2)-2*c*x1);
ex2: k*x2*(a-b*(x1+2*x2)-2*c*x2);
solve([ex1,ex2],[x1,x2]);
```

Study the stability of the fixed points. Show that the fixed point $(0, 0)$ is a repelling node?

Problem 8. Let

$$\mathbf{f}\begin{pmatrix} x_1 \\ x_2 \end{pmatrix} = \begin{pmatrix} 1 & 1 \\ 1 & 0 \end{pmatrix}\begin{pmatrix} x_1 \\ x_2 \end{pmatrix} \mod 1.$$

Then $\mathbf{f}$ is defined on the unit square in $\mathbb{R}^2$ (or on the torus). Find the Liapunov exponents of any orbit of the map. Note that the eigenvalues of the matrix are $(1 \pm \sqrt{5})/2$.

Problem 9. Let a, b, c, d be real positive parameters. The *Tinkerbell map* is $\mathbf{f} : \mathbb{R}^2 \to \mathbb{R}^2$ is given by

$$f_1(x_1, x_2) = x_1^2 - x_2^2 + ax_1 + bx_2, \qquad f_2(x_1, x_2) = 2x_1x_2 + cx_1 + dx_2$$

or written as difference equation

$$x_{1,t+1} = x_{1,t} - x_{2,t} + ax_{1,t} + bx_{2,t}, \qquad x_{2,t+1} = 2x_{1,t}x_{2,t} + cx_{1,t} + dx_{2,t}$$

where $t = 0, 1, \ldots$. Is the differential two form $dx_1 \wedge dx_2$ invariant under the map? Find the fixed points and study their stability. Calculate the first iterate of the map and their fixed points, i.e. find periodic points. Study the stability of these periodic points. Consider the parameter values $c_1 = -0.3$, $c_2 = -0.6$, $c_3 = 2$, $c_4 = 0.5$.

Problem 10. Let $r > 0$. Study the two-dimensional map

$$x_{1,t+1} = rx_{2,t}^2, \qquad x_{2,t+1} = \frac{1}{r}x_{1,t}^2$$

for $r = 2$. First find the fixed points. One of the fixed points is $(0, 0)$. Are the fixed points stable? Study the four cases with the initial values (i) $x_{1,0} = 1$, $x_{2,0} = 1$; (ii) $x_{1,0} = 1/2$, $x_{2,0} = 1$; (iii) $x_{1,0} = 1$, $x_{2,0} = 1/2$; (iv) $x_{1,0} = 1/2$, $x_{2,0} = 1/2$.

Problem 11. Consider the analytic function $\mathbf{f} : \mathbb{R}^2 \to \mathbb{R}^2$

$$f_1(x_1, x_2) = \frac{x_2}{1 + x_1^2}, \qquad f_2(x_1, x_2) = \frac{x_1}{1 + x_2^2}.$$

The function can be written as a coupled system of difference equations

$$x_{1,t+1} = \frac{x_{2,t}}{1 + x_{1,t}^2}, \qquad x_{2,t+1} = \frac{x_{1,t}}{1 + x_{2,t}^2}.$$

Find the fixed points and study their stability. Obviously $(0, 0)$ is a fixed point. Are there escaping solutions?

Problem 12. The *beam-beam map* defined on $\mathbb{R}^2$ is given by

$$x_{1,t+1} = x_{1,t}\cos(2\pi\alpha) + (x_{2,t} + 1 - e^{-x_{1,t}^2})\sin(2\pi\alpha)$$
$$x_{2,t+1} = -x_{1,t}\sin(2\pi\alpha) + (x_{2,t} + 1 - e^{-x_{1,t}^2})\cos(2\pi\alpha)$$

where α is a bifurcation parameter. Find the fixed points and study their stability.

Problem 13. Let $\mathbf{f}$ be an invertible map of $\mathbb{R}^n$ with $n \geq 2$. Let $\mathbf{p}$ be a fixed point saddle. A point that is in both the stable and unstable manifold of $\mathbf{p}$ and that is distinct from $\mathbf{p}$ is called a *homoclinic point.* If $\mathbf{x}$ is a homoclinic point, then

$$\mathbf{f}^{(t)}(\mathbf{x}) \to \mathbf{p} \quad \text{and} \quad \mathbf{f}^{(-t)}(\mathbf{x}) \to \mathbf{p}$$

as $t \to \infty$. The orbit of a homoclinic point is called a *homoclinic orbit.* A point in the stable manifold of a fixed point $\mathbf{p}$ and in the unstable manifold of a different fixed point $\mathbf{q}$ is called a *heteroclinic point.* The orbit of a heteroclinic point is called a *heteroclinic orbit.*
(i) Give an example of an invertible map in $\mathbb{R}^2$ which has a homoclinic orbit.
(ii) Give an example of an invertible map in $\mathbb{R}^2$ which has a heteroclinic orbit.

Problem 14. Solve the initial value problem for the nonlinear difference equation

$$x_{t+1} = x_t x_{t-1} \cdots x_1 x_0 + \sum_{t=0}^{t} x_t, \quad t = 0, 1, 2, \ldots.$$

Problem 15. (i) The logistic map

$$x_{t+1} = 4x_t(1 - x_t), \quad t = 0, 1, 2, \ldots$$

and $x_0 \in [0, 1]$ is probably the most studied map with chaotic behaviour. Study the two-dimensional map

$$x_{1,t+1} = 4x_{1,t}(1 - x_{1,t}), \qquad x_{2,t+1} = 4x_{1,t}x_{2,t}(1 - x_{2,t})$$

where $x_{1,0}, x_{2,0} \in [0, 1]$. This means the second equation is modulated by the solution of the logistic map. Study the higher-dimensional case

$$x_{1,t+1} = 4x_{1,t}(1 - x_{1,t})$$

$$x_{2,t+1} = 4x_{1,t}x_{2,t}(1 - x_{2,t})$$
$$x_{3,t+1} = 4x_{1,t}x_{2,t}x_{3,t}(1 - x_{3,t})$$

where $x_{1,0}, x_{2,0}, x_{3,0} \in [0, 1]$. Extend to n dimensions.

Problem 16. The map $f : [0,1] \to [0,1]$, $f(x) = \sin(2\pi x)$ or written as difference equation $x_{t+1} = \sin(2\pi x_t)$ with $x_0 \in [0,1]$ shows chaotic behaviour. Does the two-dimensional map $\mathbf{f} : [0,1] \times [0,1] \to [0,1] \times [0,1]$

$$f_1(x_1, x_2) = \sin(2\pi x_2), \qquad f_2(x_1, x_2) = \sin(2\pi x_1)$$

show chaotic behaviour? A fixed point is given by $(0,0)$. Also study the case

$$f_1(x_1, x_2) = \sin(2\pi x_2), \qquad f_2(x_1, x_2) = -\sin(2\pi x_1).$$

Problem 17. Consider the *standard map* given by

$$r_{t+1} = r_t - \frac{k}{2\pi}\sin(2\pi\theta_t), \qquad \theta_{t+1} = \theta_t + r_{t+1} \mod 1.$$

Find the fixed points in dependence of k and study their stability.

Problem 18. Study the two-dimensional map $(t = 0, 1, \ldots)$

$$\theta_{t+1} = \theta_t + r \bmod 1, \qquad x_{t+1} = x_t f(\theta_t)$$

where

$$f(\theta) := \begin{cases} -1 \text{ for } 0 \le \theta < 0.5 \\ \;\;1 \text{ for } 0.5 \le \theta < 1 \end{cases}$$

and the number r is chosen irrational. This is a skew-product dynamical system, where a variable θ satisfies a self-contained difference equation and the variable is utilized to force a second difference equation.

Problem 19. Study the two-dimensional map

$$\theta_{t+1} = \theta_t + c \mod 1, \quad x_{t+1} = f(x_t) + \epsilon\sin(2\pi\theta_t)$$

where $f(x) = 4x(1-x)$.

Problem 20. Let $a > 0$ and $b > 0$. Consider the Hénon map

$$x_{1,t+1} = a - bx_{2,t} - x_{1,t}^2, \qquad x_{2,t+1} = x_{1,t}$$

with $t = 0, 1, \ldots$. Let $|x_{1,0}| < R$, $|x_{2,0}| < R$, where R is the larger root of the characteristic equation

$$\lambda^2 - (|b| + 1)\lambda - a = 0.$$

Show that all points $(x_{1,0}, x_{2,0})$ outside of this domain tend to $+\infty$ or $-\infty$ for $t \to \infty$ or $t \to -\infty$.

Problem 21. The *skinny Baker map* is given by $B(x_1, x_2) : [0,1]\times[0,1] \to [0,1] \times [0,1]$

$$B(x_1, x_2) := \begin{cases} (x_1/3, 2x_2) & \text{if } 0 \le x_2 \le 1/2 \\ (x_1/3 + 2/3, 2x_2 - 1) & \text{if } 1/2 < x_2 \le 1. \end{cases}$$

Give a *Markov partition* of the skinny Baker map.

Problem 22. Let $x_{1,0}, x_{2,0} \in [-1, 1]$. Study the system of difference equations

$$x_{1,t+1} = 1 - 2(|x_{2,t}|)^{1/2}, \qquad x_{2,t+1} = 1 - 2(|x_{1,t}|)^{1/2}.$$

Problem 23. A delayed version of the *circle map* is given by

$$\theta_{t+1} = \theta_t + r_1 \sin(2\pi\phi_t) + r_2, \quad \phi_{t+1} = \theta_t$$

where $t = 0, 1, \ldots$ and r_1, r_2 are positive bifurcation parameters. Show that at its parameter plane (r_1, r_2) the map has a symmetry line $r_2 = 0.5$ and various Arnold tongues.

Problem 24. Let $f : \mathbb{R} \to \mathbb{R}$ be a smooth function which admits at least one root, i.e. $f(x) = 0$ admits a solution. The *regular falsi method* for finding the roots of f is given by the second order nonlinear difference equation

$$x_{t+2} = \frac{x_t f(x_{t+1}) - x_{t+1} f(x_t)}{f(x_{t+1}) - f(x_t)}, \quad t = 0, 1, \ldots$$

(i) Show that with $y_{t+1} = x_{t+2}$ the second order difference equation can be written as

$$x_{t+1} = y_t$$
$$y_{t+1} = \frac{x_t f(y_t) - y_t f(x_t)}{f(y_t) - f(x_t)}.$$

Study these equations for the logistic map $f(x) = 4x(1-x)$. Select the initial values $(x_0 = 1/2, x_1 = 1/4)$ and $(x_0 = 1/2, x_1 = 3/4)$.

Problem 25. The map $g : [-1, 1] \to [-1, 1]$, $g(x) = 1 - 2x^2$ shows fully developed chaos. The fixed points of g are $x_1^* = -1$ and $x_2^* = 1/2$ and are unstable. Let $r_1 \in (0,2]$, $r_2 > 0$ be the bifurcation parameters. Consider the smooth map $\mathbf{f} : \mathbb{R}^2 \to \mathbb{R}^2$ given by

$$f_{1,r_1,r_2}(x_1,x_2) = 1 - r_1 x_1^2 + r_2(x_2 - x_1)$$
$$f_{2,r_1,r_2}(x_1,x_2) = 1 - r_1 x_2^2 + r_2(x_1 - x_2).$$

(i) Find the fixed points of the map and study their stability.
(ii) The second iterate is given by

$$f_{1,r_1,r_2}(f_{1,r_1,r_2}(x_1,x_2), f_{2,r_1,r_2}(x_1,x_2))$$
$$= 1 - r_1(1 - r_1 x_1^2 + r_2(x_2 - x_1))^2 + r_2(x_1 - x_2)(r_1(x_1 + x_2) + 2r_2)$$
$$f_{2,r_1,r_2}(f_{1,r_1,r_2}(x_1,x_2), f_{2,r_1,r_2}(x_1,x_2))$$
$$= 1 - r_1(1 - r_1 x_2^2 + r_2(x_1 - x_2))^2 + r_2(x_2 - x_1)(r_1(x_1 + x_2) + 2r_2).$$

Find the fixed points of the second iterate (and thus periodic orbits) and study their stability.

Problem 26. Let $r_1, r_2 > 1$ and $a_1, a_2 > 0$. Study the prey-predator model

$$x_{1,t+1} = r_1 x_{1,t}(1 - x_{1,t})(1 - a_1 x_{2,t})$$
$$x_{2,t+1} = r_2 x_{2,t}(1 - x_{2,t})(1 - a_2(1 - x_{1,t})).$$

Problem 27. A *predator-prey model* is described by the two-dimensional map

$$x_{1,t+1} = x_{1,t}\exp(b(1 - x_{1,t}/K) - a x_{2,t}), \quad x_{2,t+1} = x_{1,t}(1 - \exp(-a x_{2,t}))$$

where $K, a, b > 0$. Here x_1 denotes the prey and x_2 denotes the predator. Find the fixed points and study their stability.

Problem 28. Study the two-dimensional map

$$x_{1,t+1} = \frac{1}{2}\ln(\cosh(4x_{2,t})), \quad x_{2,t+1} = \frac{1}{2}\ln(\cosh(4x_{1,t}))$$

with $t = 0, 1, \ldots$ and the initial conditions $x_{1,0} = 1/2$ and $x_{2,0} = 1$.

Problem 29. Study the two-dimensional map

$$x_{1,t+1} = \sin(x_{1,t} + x_{2,t}), \qquad x_{2,t+1} = \cos(x_{1,t} - x_{2,t}).$$

Problem 30. Study the second order difference equation

$$x_{t+1} + x_{t-1} = -x_t + \frac{at + b}{x_t} + c$$

where $t = 0, 1, \ldots$ and $x_{-1} = 0$, $x_0 = 1/2$, $a = 1$, $b = 1$, $c = 1$.

Problem 31. Study the two-dimensional map

$$x_{1,t+1} = e^{x_{2,t}} x_{1,t}, \qquad x_{2,t+1} = e^{-x_{1,t}} x_{2,t}$$

with $t = 0, 1, \ldots$ and $x_{1,0} = 1$, $x_{2,0} = 1$. The fixed point of the map is $(0, 0)$. Is the fixed point stable?

Problem 32. Consider the two-dimensional map $\mathbf{f} : \mathbb{R}^2 \to \mathbb{R}^2$

$$f_1(x_1, x_2) = -x_2 + g(x_1), \qquad f_2(x_1, x_2) = x_1$$

or written as a system of difference equations

$$x_{1,t+1} = -x_{2,t} + g(x_{1,t}), \qquad x_{2,t+1} = x_{1,t}$$

where $t = 0, 1, \ldots$ and $g : \mathbb{R} \to \mathbb{R}$ is the non-invertible map $g(x) = 4x(1-x)$. Find the fixed points of the map and study their stability. Find the first iterate and study the stability of the periodic points.

Problem 33. Study the coupled system of first order difference equations

$$m(t+1) = \tanh\left(\frac{p}{T}m(t) - \frac{1}{T^2}q(t)\right)$$
$$q(t+1) = m(t)\text{sech}^2\left(\frac{p}{T}m(t) - \frac{1}{T^2}q(t)\right)$$

where $= 0, 1, 2, \ldots$ and the bifurcation parameter are in the range $-3 < p < 3$, $0 < T < 2$. Find the fixed points. Find periodic orbits. Does the system show chaotic behaviour? This model plays a role for an an Ising model on a regular Cayley tree, with competing ferro- and antiferromagnetic nearest-neighbour interactions.

Problem 34. Consider the 18-parameter family of mappings of the plane given by

$$x' = \frac{f_1(y) - xf_2(y)}{f_2(y) - xf_3(y)}, \qquad y' = \frac{g_1(x') - yg_2(x')}{g_2(x') - yg_3(x')} \tag{1}$$

with

$$\mathbf{f}(x) := (\mathbf{A}_0\mathbf{X}) \times (\mathbf{A}_1\mathbf{X}), \qquad \mathbf{g}(x) := (\mathbf{A}_0^T\mathbf{X}) \times (\mathbf{A}_1^T\mathbf{X}) \tag{2a}$$

where $\times$ denotes the vector product and

$$\mathbf{X} := \begin{pmatrix} x^2 \\ x \\ 1 \end{pmatrix}, \qquad \mathbf{A}_i := \begin{pmatrix} \alpha_i \, \beta_i \, \gamma_i \\ \delta_i \, \epsilon_i \, \xi_i \\ \kappa_i \, \lambda_i \, \mu_i \end{pmatrix}, \qquad i = 0, 1. \tag{2b}$$

Show that each member of this family possesses a 1-parameter family of *invariant curves* that fills the plane

$$(\alpha_0+K\alpha_1)x^2y^2+(\beta_0+K\beta_1)x^2y+(\gamma_0+K\gamma_i)x^2+(\delta_0+K\delta_1)xy^2+(\epsilon_0+K\epsilon_1)xy$$
$$+(\xi_0+K\xi_1)x+(\kappa_0+K\kappa_1)y^2+(\lambda_0+K\lambda_1)y+(\mu_0+K\mu_1)=0$$

where the integration constant K is invariant on each curve.

Problem 35. (i) Study the difference equation ($t=0,1,2,\ldots$)

$$\det\begin{pmatrix} x_{t+2} & x_t \\ x_t & x_{t+1} \end{pmatrix}=0$$

with $x_0=1$, $x_1=1$.
(ii) Consider the difference equation given by

$$\det\begin{pmatrix} x_{t+1} & x_t \\ x_t & x_{t-1} \end{pmatrix}=1$$

with $t=1,2,\ldots$. Find the solution with $x_0=x_1=1$. Find the solution with $x_0=x_1=1$ and $x_t \bmod 2$.
(iii) Study the initial value problem of the difference equation

$$\det\begin{pmatrix} x_{t+3} & x_{t+2} & x_{t+1} \\ x_{t+2} & x_{t+1} & x_t \\ x_{t+1} & x_t & 0 \end{pmatrix}=0$$

where $t=0,1,2,\ldots$ and $x_0=1$, $x_1=1/2$, $x_2=1/3$.

Problem 36. Study the map

$$f_1(x,\phi)=\frac{1}{2}(e^{2x}\sin^2(\phi)+e^{-2x}\cos^2(\phi))$$
$$f_2(x,\phi)=\frac{1}{2}(e^{2x}\cos^2(\phi)+e^{-2x}\sin^2(\phi)).$$

First find the fixed points. Note that $f_1(x,\phi)+f_2(x,\phi)=\cosh(2x)$.

Problem 37. Let $R>r>0$. Study the map $\mathbf{f}:\mathbb{R}^2\to\mathbb{R}^2$

$$f_1(x_1,x_2)=(x_1 \quad 1-x_1)\begin{pmatrix} R & 0 \\ 0 & r \end{pmatrix}\begin{pmatrix} x_2 \\ 1-x_2 \end{pmatrix}$$
$$=((R+r)x_2-r)x_1+r(1-x_2)$$

$$f_2(x_1,x_2)=(x_1 \quad 1-x_1)\begin{pmatrix} r & 0 \\ 0 & R \end{pmatrix}\begin{pmatrix} x_2 \\ 1-x_2 \end{pmatrix}$$
$$=((R+r)x_1-R)x_2+R(1-x_1).$$

First find the fixed points and study their stability.

Problem 38. In a saddle-node bifurcation a pair of periodic orbits are created "out of nothing". One of the periodic orbits is always unstable (the saddle) while the other periodic orbit is always stable (the node). Give a two-dimensional map that shows a saddle-node bifurcation.

Problem 39. Study the two-dimensional map

$$x_{1,t+1} = 4x_{1,t}(1 - x_{1,t}), \qquad x_{2,t+1} = 2x_{1,t}x_{2,t} \bmod 1$$

where $t = 0, 1, 2, \ldots$ and $x_{1,0}, x_{2,0} \in [0, 1]$.

Problem 40. Study the modulated *circle map*

$$\theta_{t+1} = \theta_t + a_1 \sin(2\pi\theta_t) + a_2 \sin(2\pi\phi_t) + r, \quad \text{mod } 1$$
$$\phi_{t+1} = \phi_t + b$$

where $a_1 = 0.15$, $a_2 = 0.01$, $b = (\sqrt{5} - 1)/2$ and r is the bifurcation parameter $r \in (0, 1)$.

Problem 41. Consider the two equations

$$f_1(x_1, x_2) = 0, \qquad f_2(x_1, x_2) = 0$$

with $f_1(x_1, x_2) = x_1^2 + x_2^2 - 1$ and $f_2(x_1, x_2) = x_1 + x_2$. Study the coupled system of difference equations

$$x_{1,t+1} = x_{1,t} - \frac{1}{f_{1,1}(x_{1,t}, x_{2,t})} f_1(x_{1,t}, x_{2,t})$$
$$x_{2,t+1} = x_{2,t} - \frac{1}{2} \frac{1}{f_{2,2}(x_{1,t+1}, x_{2,t})} f_2(x_{1,t+1}, x_{2,t})$$

where $t = 0, 1, 2, \ldots$, $f_{1,1}$ is the partial derivative of f_1 with respect to x_1 and $f_{2,2}$ is the partial derivative of f_2 with respect to x_2. Consider the initial values $x_{1,0} = 0.2$, $x_{2,0} = 0.8$.

Problem 42. Study the coupled *circle map*

$$\theta_{t+1} = \theta_t + a_1 \sin(2\pi\theta_t) + a_2 \sin(2\pi\phi_t) + r_1, \quad \text{mod } 1$$
$$\phi_{t+1} = \phi_t + b_1 \sin(2\pi\phi_t) + b_2 \sin(2\pi\theta_1) + r_2, \quad \text{mod } r_2$$

with $a_1 = a_2 = b_1 = b_2 = 1/2$ and $r_1, r_2 \in (0, 1)$ be the bifurcation parameters.

Problem 43. Let n be a positive integer with $n \geq 2$. Let c be any real number. We define x_t recursively by $x_0 = 0$, $x_1 = 1$ and for $t \geq 0$,

$$x_{t+2} = \frac{cx_{t+1} - (n-t)x_t}{t+1}.$$

Fix n and then take c to be the largest value for which $x_{t+1} = 0$. Find x_t in terms of n and t, $2 \leq t \leq n$.

Problem 44. Study the map $\mathbf{f} : \mathbb{R}^2 \to \mathbb{R}^2$

$$f_1(x_1, x_2) = e^{x_1^2 - x_2^2} \cos(2x_1x_2), \qquad f_2(x_1, x_2) = e^{x_1^2 - x_2^2} \sin(2x_1x_2).$$

First find the fixed points if there any. Is the map invertible?

Problem 45. Consider the map $\mathbf{f} : \mathbb{N}_0 \times \mathbb{N}_0 \to \mathbb{N}_0 \times \mathbb{N}_0$ given by

$$f_1(n_1, n_2) = \frac{1}{2}((n_1 + n_2)^2 + n_1 + 3n_2)$$

$$f_2(n_1, n_2) = \frac{1}{2}((n_1 + n_2)^2 + 3n_1 + n_2).$$

Show that $(0, 0)$ is a fixed point. Are there other fixed points? Is the map invertible?

Problem 46. Let $r > 0$ be the bifurcation parameter. Study the coupled system of maps

$$x_{1,t+1} = r\left(x_{1,t} - \frac{1}{4}(x_{1,t} + x_{2,t})^2\right), \quad x_{2,t+1} = \frac{1}{r}\left(x_{2,t} + \frac{1}{4}(x_{1,t} + x_{2,t})^2\right).$$

Obviously $(0, 0)$ is a fixed point. Is the fixed point stable?

Problem 47. Let $A = (a_{jk})$ be a given 3×3 matrix over $\mathbb{R}$ with $\det(A) \neq 0$. Is the transformation

$$f_1(x_1, x_2) = \frac{a_{11}x_1 + a_{12}x_2 + a_{13}}{a_{31}x_1 + a_{32}x_2 + a_{33}}$$

$$f_2(x_1, x_2) = \frac{a_{21}x_1 + a_{22}x_2 + a_{23}}{a_{31}x_1 + a_{32}x_2 + a_{33}}$$

invertible? If so find the inverse. Find the fixed points.

Problem 48. Study the *Gingerbreadman map* $\mathbf{f} : \mathbb{R}^2 \to \mathbb{R}^2$

$$f_1(x_1, x_2) = 1 - x_2 + |x_1|, \quad f_2(x_1, x_2) = x_1$$

or written as system of difference equations

$$x_{1,t+1} = 1 - x_{2,t} + |x_{1,t}|, \qquad x_{2,t+1} = x_{1,t} \quad t = 0, 1, \ldots$$

First find fixed points and periodic points and study their stability. Is the map invertible?

Problem 49. Study the *annulus map*

$$\Theta_{t+1} = \Theta_t + \Omega + r_{t+1}, \quad r_{t+1} = br_t - \frac{k}{2\pi}\sin(2\pi\Theta_t)$$

where k, b, Ω are the non-negative bifurcation parameters. For $b = 0$ the map reduces to the sine-circle map and for $b = 1$ we have the standard map.

Problem 50. According to Gauss, the *elliptic integral*

$$I = \frac{2}{\pi}\int_0^{\pi/2} \frac{dx}{(a^2\cos^2(x) + b^2\sin^2(x))^{1/2}}$$

is equal to the limit of any of the two convergent sequences

$$s_0, s_1, s_2, \ldots \quad \text{or} \quad t_0, t_1, t_2 \ldots$$

as defined by the recurrence relations for $j > 0$

$$s_{j+1} = (s_j + t_j)/2, \quad t_{j+1} = \sqrt{s_j t_j}$$

and $s_0 = a$, $t_0 = b$. The calculation of the two sequences is called the arithmetic-geometric mean method. Write a C++ program that implements this method.

Problem 51. Consider the Lie group $SL(2, \mathbb{R})$, i.e. the set of all real 2×2 matrices with determinant equal to 1. A dynamical system in $SL(2, \mathbb{R})$ can be defined by

$$M_{k+2} = M_k M_{k+1}, \qquad k = 0, 1, 2, \ldots$$

with the initial matrices $M_0, M_1 \in SL(2, \mathbb{R})$. Let $F_k := \mathrm{tr}(M_k)$. Is

$$F_{k+3} = F_{k+2}F_{k+1} - F_k \qquad k = 0, 1, 2, \ldots ?$$

Prove or disprove. Use the property that for any 2×2 matrix A we have (*Cayley-Hamilton theorem*)

$$A^2 - A\mathrm{tr}(A) + I_2 \det(A) = 0_2.$$

Problem 52. Let $a > 0$, $b > 0$. The *Duffing map* $\mathbf{f} : \mathbb{R}^2 \to \mathbb{R}^2$ is given by

$$f_1(x_1, x_2) = x_2, \qquad f_2(x_1, x_2) = -bx_1 + ax_2 - x_2^3$$

or written as difference equation

$$x_{1,t+1} = x_{2,t}, \qquad x_{2,t+1} = -bx_{1,t} + ax_{2,t} - x_{2,t}^3$$

where $t = 0, 1, \ldots$. Is $dx_1 \wedge dx_2$ invariant under the map? Find the fixed points and study their stability. Calculate the first iterate of the map and their fixed points, i.e. find periodic points. Study the stability of these periodic points.

Problem 53. The recurrence relation

$$x_{t+2} = x_{t+1} + x_t, \qquad t = 0, 1, \ldots$$

with $x_0 = x_1 = 1$ provides the *Fibonacci sequence.*
(i) Study the recurrence relation

$$x_{t+2} = e^{i\pi t} x_{t+1} + x_t.$$

Note that

$$e^{i\pi t} \equiv \cos(\pi t) + i \sin(\pi t) \equiv \cos(\pi t)$$

with $\cos(\pi t) = 1$ if t is even and $\cos(\pi t) = -1$ if t is odd.
(ii) Study the recurrence relation

$$x_{t+2} = e^{i\pi x_t} x_{t+1} + x_t$$

with $t = 0, 1, \ldots$ and $x_0 = x_1 = 1$.
(iii) Study the case $(t = 0, 1, \ldots)$

$$x_{t+2} = x_{t+1} + x_t \mod 2$$

with $x_0 = 0$, $x_1 = 1$. Is the sequence eventually periodic?
(iv) Study the map

$$x_{t+1} = (x_t + x_{t-1}) \bmod p \quad t = 1, 2, \ldots$$

for $p = 7$ and $x_0 = 1$, $x_1 = 2$.
(v) Study the recursion

$$\theta_{t+1} = 2\theta_t + \theta_{t-1} \mod 2, \quad t = 1, 2, \ldots$$

with the initial conditions (a) $\theta_0 = 0$, $\theta_1 = 0$; (b) $\theta_0 = 1$, $\theta_1 = 1$; (c) $\theta_0 = 0$, $\theta_1 = 1$; (d) $\theta_0 = 1$, $\theta_1 = 1$.

Problem 54. Consider the skew-tent map $f : [0,1] \to [0,1]$ given by

$$f(x) = \begin{cases} x/r & \text{for } 0 \le x \le r \\ (1-x)/(1-r) & \text{for } r < x \le 1 \end{cases}$$

where $0.5 \le r < 1$. Consider the coupled system

$$x_{1,t+1} = f(x_{1,t} + \delta(x_{2,t} - x_{1,t})), \qquad x_{2,t+1} = f(x_{2,t} + \epsilon(x_{1,t} - x_{2,t}))$$

where $t = 0, 1, \ldots$. Let $\delta = 0$, $\epsilon = 1$ and $x_{1,0} = 0.2$, $x_{2,0} = 0.3$ (master-slave system). Does the system synchronizes?

Problem 55. Study the extended *Arnold cat map*

$$\begin{pmatrix} x_1 \\ x_2 \\ x_3 \\ x_4 \end{pmatrix} \mapsto \left(\begin{pmatrix} 1 & 1 \\ 1 & 2 \end{pmatrix} \otimes \begin{pmatrix} 1 & 1 \\ 1 & 2 \end{pmatrix} \right) \begin{pmatrix} x_1 \\ x_2 \\ x_3 \\ x_4 \end{pmatrix} \quad \text{mod } 1$$

where $\otimes$ denotes the Kronecker product.

Problem 56. Study the one-parameter family of maps $\mathbf{f} : \mathbb{R}^2 \to \mathbb{R}^2$ defined by

$$\mathbf{f} : \begin{pmatrix} x_{1,n+1} \\ x_{2,n+1} \end{pmatrix} = \begin{cases} \begin{pmatrix} \eta & 0 \\ 0 & \eta^{-1} \end{pmatrix} \begin{pmatrix} x_{1,n} \\ x_{2,n} \end{pmatrix} & \text{for } l(x_{1,n}, x_{2,n}) \le 0 \\ \begin{pmatrix} \eta & 0 \\ 0 & \eta^{-1} \end{pmatrix} \begin{pmatrix} x_{1,n} \\ x_{2,n} \end{pmatrix} + (1-\eta) \begin{pmatrix} 1 \\ -\eta^{-1} \end{pmatrix} & \text{for } l(x_{1,n}, x_{2,n}) > 0 \end{cases}$$

where $l(x_1, x_2)$ is defined as

$$C : \; l(x_1, x_2) := \eta x_1 + x_2 - \frac{1+\eta}{2}.$$

It separates the two branches of $\mathbf{f}$. The behaviour of $\mathbf{f}$ depends on the parameter η.

Problem 57. Let $a > 0$ and $b > 0$. The *Lozi map* $\mathbf{f} : \mathbb{R}^2 \to \mathbb{R}^2$ is given by

$$f_1(x_1, x_2) = 1 + x_2 - a|x_1|, \qquad f_2(x_1, x_2) = bx_1.$$

Show that if $b \in (0,1)$, $a > 0$, $2a + b < 4$, $b < (a^2 - a)/(2a+1)$ and $\sqrt{2}a > b + 2$ then there is a hyperbolic fixed point of saddle type. The fixed point is given by

$$x_1^* = (a + 1 - b)^{-1}, \quad x_2^* = bx_1^*.$$

Problem 58. Let $a > 0$, $d > 0$. Consider the coupled logistic map $\mathbf{f} : \mathbb{R}^2 \to \mathbb{R}^2$

$$f_1(x_1, x_2) = 1 - ax_1^2 + d(x_2 - x_1), \qquad f_2(x_1, x_2) = 1 - ax_2^2 + d(x_1 - x_2)$$

or written as difference equation

$$x_{1,t+1} = 1 - ax_{1,t}^2 + d(x_{2,t} - x_{1,t}), \qquad x_{2,t+1} = 1 - ax_{2,t}^2 + d(x_{1,t} - x_{2,t}).$$

(i) Is $dx_1 \wedge dx_2$ invariant under the map?
(ii) Find the fixed points of the map and study their stability.
(iii) Calculate the first iterate of the map and their fixed points, i.e. find periodic points. Study the stability of these periodic points.
(iv) Apply the transformation

$$v_{1,t} = \frac{1}{2}(x_{1,t} + x_{2,t}), \qquad v_{2,t} = \frac{1}{2}(x_{1,t} - x_{2,t})$$

and show that $v_{2,t}$ is given by

$$v_{2,t+1} = \prod_{\tau=0}^{t+1}(-2(av_{1,\tau} + d))v_{2,0}.$$

(v) Show that the stability of the $x_{1,t} = x_{2,t} = v_{1,t}$ is given by

$$\left| \prod_{\tau=1}^{p}(-2(ax_{1,t} + d)) \right| < 1.$$

(vi) Let $d = 0.1$. Show that *Hopf bifurcation* occurs at $a \geq 1$.

Problem 59. Let $a > 0$ and $1 > b > 0$. Consider the system of nonlinear two-dimensional difference equations

$$x_{1,t+1} = x_{1,t} + a(e^{x_{2,t} - x_{1,t}} - 1), \; x_{2,t+1} = x_{2,t} + b(x_{1,t} - x_{2,t}), \; t = 0, 1, 2, \ldots.$$

(i) Find the fixed points. Find the variational equation and study the stability of the fixed points. Solving the system of equations

$$x_1^* + a(e^{x_2^* - x_1^*} - 1) = x_1^*, \quad x_2^* + b(x_1^* - x_2^*) = x_2^*$$

provides the solution $x_1^* = x_2^*$ for the fixed points.
(ii) Find the second iterate.
(iii) Show that if $x_{2,t} > x_{1,t}$, then $x_{1,t+1}$ increases and $x_{2,t+1}$ decreases.
(iv) Show that

$$x_{2,t} = b \sum_{j=1}^{\infty}(1 - b)^{j-1} x_{1,t-j}.$$

Problem 60. Let $r_1 > 0$ and $r_2 > 0$. Study the two-dimensional map

$$x_{1,t+1} = r_1 x_{1,t}(1 - x_{1,t} - x_{2,t}), \quad x_{2,t+1} = r_2 x_{1,t} x_{2,t}.$$

First find the fixed points and study their stability.

Problem 61. Study the two-dimensional map

$$x_{1,t+1} = 4x_{1,t}(1 - x_{1,t}), \quad x_{2,t+1} = x_{1,t} x_{2,t}, \quad t = 0, 1, 2, \ldots$$

The fixed points follow from the solution of the equations

$$4x_1^*(1 - x_1^*) = x_1^*, \quad x_1^* x_2^* = x_2^*$$

as $(x_1^*, x_2^*) = (0, 0)$ and $(x_1, x_2) = (3/4, 0)$

Problem 62. Let $a > 0$, $b > 0$. Consider the map $\mathbf{f} : \mathbb{R}^2 \to \mathbb{R}^2$

$$f_1(x_1, x_2) = a - bx_1 + x_1^2 x_2, \quad f_2(x_1, x_2) = bx_1 + x_2 - x_1^2 x_2.$$

Let $a = 1$ and consider b as the bifurcation parameter. The fixed points are $x_1^* = 1$, $x_2^* = b$.

Problem 63. Show that the *McMillan maps*

$$f_1(x_1, x_2) = x_2, \quad f_2(x_1, x_2) = -x_1 - \frac{\beta x_2^2 + \epsilon x_2 + \xi}{\alpha x_2^2 + \beta x_2 + \gamma}$$

are a family of area-preserving rational maps preserving the biquadratic foliation

$$\alpha x_1^2 x_2^2 + \beta(x_1^2 x_2 + x_1 x_2^2) + \gamma(x_1^2 + x_2^2) + \epsilon x_1 x_2 + \xi(x_1 + x_2) + K = 0$$

where K is the parameter which parametrizes each invariant curve in the plane. This a typical problem for computer algebra.

2.3 Complex Maps

Let $\mathbb{C}$ be the complex numbers. A complex number z can be written as $z = x + iy$, where $x, y \in \mathbb{R}$ and $z = re^{i\phi}$ with $r \geq 0$. The complex conjugate of z is $\bar{z} = x - iy$ and $\bar{z} = re^{-i\phi}$. Thus $z\bar{z} = r^2$. Furthermore $|z| = \sqrt{x^2 + y^2}$ and $|z| = r$.

A complex map is a map $f : \mathbb{C} \to \mathbb{C}$. In particular important are *analytic maps* such as

$$f(z) = 2z + z^2, \quad f(z) = \exp(z), \quad f(z) = \sin(2z), \quad f(z) = i + iz.$$

Sometimes one has to work with the extended complex plane $\bar{\mathbb{C}} := \mathbb{C} \cup \{\infty\}$ (also called the *Riemann sphere*) and adopt the convention that $1/0 = \infty$ and $1/\infty = 0$.

Let $f : \mathbb{C} \to \bar{\mathbb{C}}$ be a transcendental meromorphic function. Consider the iteration

$$z_{t+1} = f(z_t), \quad t = 0, 1, 2, \ldots$$

The complex number z^* is called a fixed point of f if $f(z^*) = z^*$. A fixed point of f is *attractive* if

$$|df(z = z^*)/dz| < 1.$$

A fixed point of f is called *superattracting* if

$$df(z = z^*)/dz = 0.$$

A fixed point of f is *repulsive* if

$$|df(z = z^*)/dz| > 1.$$

A fixed point of f is *indifferent* if

$$|df(z = z^*)/dz| = 1.$$

With $z = x + iy$ $(x, y \in \mathbb{R})$ we have

$$\begin{aligned}
e^z &= e^x e^{iy} = e^x(\cos(y) + i\sin(y)) \\
\sin(z) &= \sin(x + iy) = \cosh(y)\sin(x) + i\sinh(y)\cos(x) \\
\cos(z) &= \cos(x + iy) = \cos(x)\cosh(y) - i\sin(x)\sinh(y) \\
\sinh(z) &= \sinh(x + iy) = \sinh(x)\cos(y) + i\cosh(x)\sin(y) \\
\cosh(z) &= \cosh(x + iy) = \cosh(x)\cos(y) + i\sinh(x)\sin(y).
\end{aligned}$$

2.3.1 Solved Problems

Problem 1. Study the behaviour of the fixed points of the complex map $f : \mathbb{C} \to \mathbb{C}$

$$f(z) = z^2 \tag{1}$$

i.e. find the fixed point from $f(z^*) = z^*$ and study their stability.

Solution 1. The only fixed points of f are 0 and 1, which follow from the solution of $f(z^*) = z^*$. Since

$$\frac{df(z^* = 0)}{dz} = 0 \tag{2}$$

we find that 0 is an attracting fixed point. Since

$$\frac{df(z^* = 1)}{dz} = 2 \tag{3}$$

we find that 1 is a repelling fixed point. We set

$$z = r\exp(i\theta) \tag{4}$$

where $r \geq 0$. Then

$$f(z) = r^2 \exp(2i\theta) \tag{5}$$

In general we find

$$f^{(n)}(z) = r^{2^n} \exp(2^n i\theta).$$

Since $|f^{(n)}(z)| = r^{2^n}$, we find that z is in $W^s(0)$ if and only if $|z| < 1$ and z is in $W^s(\infty)$ if and only if $|z| < 1$. Consider now the dynamics for $|z| = 1$. Note that $|z| = 1$ implies $|f^{(n)}(z)| = 1$ for all n. If we denote the circle of points whose modulus is one by S^1, then we can write $f : S^1 \to S^1$. If $\arg(z) = \theta$ then $\arg(f(z)) = 2\theta$. This is the doubling map on the circle which is chaotic.

Problem 2. Let

$$f(z) = \frac{az + b}{cz + d}, \qquad ad - bc \neq 0 \tag{1}$$

be a *Möbius transformation.* The function f is defined and continuous on the extended complex plane and

$$f : \mathbb{C} \cup \{\infty\} \to \mathbb{C} \cup \{\infty\} \tag{2}$$

is one-to-one and onto. f is a homeomorphism of the extended complex plane.

(i) Find the inverse of f.
(ii) Find the derivative of f.
(iii) Find the fixed points of

$$f(z) = \frac{az+b}{cz+d}$$

with $ad - bc = 1$ and $c \neq 0$.

Solution 2. (i) The inverse function is given by

$$f^{-1}(z) = \frac{dz-b}{-cz+a}.$$

(ii) We have

$$\frac{df(z)}{dz} = \frac{ad-bc}{(cz+d)^2}.$$

If $ad - bc = 1$ we have

$$\frac{df(z)}{dz} = \frac{1}{(cz+d)^2}.$$

(iii) The fixed point equation

$$\frac{az+b}{cz+d} = z$$

provides the quadratic equation

$$z^2 + \frac{d-a}{c}z - \frac{b}{c} = 0$$

with the solutions

$$z = -\frac{d-a}{2c} \pm \sqrt{(a+d)^2 - 4}.$$

Thus if $a + d = 2$ or $a + d = -2$ we only have one fixed point, namely $z^* = -(d-a)/(2c)$.

Problem 3. (i) Consider the complex map $f : \mathbb{C} \to \mathbb{C}$

$$f(z) = cz$$

where c is a complex number with $|c| \neq 1$. Find the fixed points of this map. Find $f^{(n)}$ and discuss the cases $|c| < 1$ and $|c| > 1$.
(ii) The complex map $g : \mathbb{C} \to \mathbb{C}$ is defined by

$$g(z) = az + b$$

where $a, b \in \mathbb{C}$. Find for which values of a and b there exists a fixed point of g. Note that if $a \neq 1$, then g is topologically conjugate to function of the form $f(z) = cz$.

Solution 3. (i) From $f(z^*) = cz^* = z^*$ we find that 0 is the only fixed point of this map. We have for the second iterate $f^{(2)}(z) = c^2 z$ and in general for the n-th iterate

$$f^{(n)}(z) = c^n z.$$

Let $z_0 \neq 0$. Since the complex number c can be written as $c = |c|e^{i\theta}$ we have $c^n = |c|^n e^{in\theta}$. Thus we can write

$$f^{(n)}(z_0) = |c|^n e^{in\theta} z_0 = |c|^n |z_0| \exp((n\theta + \arg(z_0))i).$$

It follows that $|f^{(n)}(z_0) = |c|^n |z_0|$ and $\arg(f^{(n)}(z_0)) = \arg(z_0) + n\theta$. If $|c| < 1$ then $|c|^n |z_0|$ converges to the fixed point 0 as n tends to infinity. Thus $W^s(0) = \mathbb{C}$ when $|c| < 1$. For the case $|c| > 1$ we find that $|c|^n |z_0|$ tends to infinity. Thus $W^s(0) = 0$ and $W^s(\infty)$ is all of $\mathbb{C}$ except 0.
(ii) The fixed point equation is $az^* + b = z^*$. This means that g will have a fixed point if $b = 0$ or if $b \neq 0$, $a \neq 1$. A fixed point does not exist if $a = 1$, $b \neq 0$. Then $g(z) = z + b$. Iteration of this map yields the expression $g^{(n)}(z) = z + nb$ for the n-th iterate. This implies that $|g^{(n)}(z)| \to \infty$ as $n \to \infty$.

Problem 4. Consider the complex map $f : \mathbb{C} \to \mathbb{C}$

$$f(z) = \frac{z}{1 + z\bar{z}}.$$

Find the fixed points. Iterate $f(1)$, $f(-1)$, $f(i)$, $f(-i)$.

Solution 4. From $f(z^*) = z^*$ we find the only fixed point $z^* = 0$. Note that

$$\frac{|z|}{1 + z\bar{z}} \leq |z|$$

for all $z \in \mathbb{C}$. The sequences tend to 0.

Problem 5. Consider the complex analytic map $f_c : \mathbb{C} \to \mathbb{C}$

$$f_c(z) = z^2 + c$$

where $z, c \in \mathbb{C}$ and c fixed.
(i) Find the fixed points of f_c.
(ii) Find the fixed points of $f_c(f_c)$, i.e. periodic points of f_c.

Solution 5. (i) From the quadratic equation $(z^*)^2 + c = z^*$ we obtain the fixed points

$$z_1^* = \frac{1}{2} - \frac{1}{2}\sqrt{1-4c}, \quad z_2^* = \frac{1}{2} + \frac{1}{2}\sqrt{1-4c}.$$

Note that $z_1^* + z_2^* = 1$.
(ii) Since

$$f_c(f_c(z)) = (z^2 + c^2) + c = z^4 + 2cz^2 + c^2 + c$$

we have to solve the quartic equation $f_c(f_c(z)) = z$, i.e.

$$z^4 + 2cz^2 + c^2 + c = z.$$

This equation can be written as

$$(z^2 + c - z)(z^2 + z + c + 1) = 0.$$

From $z^2 + c - z = 0$ we obtain the fixed points given in (i). From $z^2 + z + c + 1 = 0$ we obtain the additional fixed points

$$z_3^* = -\frac{1}{2} + \sqrt{(-c - 3/4)}, \quad z_4^* = -\frac{1}{2} - \sqrt{(-c - 3/4)}$$

of $f_c(f_c)$ and thus periodic points of f_c, i.e.

$$f_c(z_3^*) = z_4^*, \qquad f_c(z_4^*) = z_3^*.$$

Problem 6. Consider the complex map $f : \mathbb{C} \to \mathbb{C}$ defined by $f(z) = z^3$.
(i) Find the fixed points of f and determine if they are attracting, repelling or non-hyperbolic.
(ii) Find all periodic points of f. Determine if they are attracting, repelling or non-hyperbolic.

Solution 6. (i) Fixed points follow from the solution of the equation $z^3 = z$, i.e.

$$z(z^2 - 1) = 0.$$

The fixed points are all real: $z_0^* = 0$ and the square roots of 1, i.e. $z_{1,2}^* = \pm 1$. Furthermore, $f'(z) = 3z$, and therefore $|f'(z_0^*)| = 0$, $|f'(z_{1,2}^*| = 3$. Hence z_0^* is attracting and $z_{1,2}^*$ are repelling.
(ii) The n-th iterate of f is given by $f^{(n)}(z) = z^{3^n}$. The periodic points of period n therefore follow from the equation $z^{3^n} = z$, i.e.

$$z\left(z^{3^n - 1} - 1\right) = 0.$$

Trivially, $z_0 = 0$ is one of the period-n points. The other $3^n - 1$ periodic points are given by the $3^n - 1$-th roots of 1

$$z_k = \exp\left(\frac{2k\pi}{3^n - 1} i\right), \quad k = 1, 2, \ldots, 3^n - 1.$$

For the derivative of the n-th iterate it follows from the chain rule that

$$\begin{aligned}(f^{(n)})'(z) &= f'(f(z))\, f'(f^{(2)}(z)) \ldots f'(f^{(n-1)}(z)) \\ &= 3^n\, f(z)\, (f^{(2)}(z)) \ldots f^{(n-1)}(z).\end{aligned}$$

Hence $(f^{(n)})'(z_0) = 0$, and z_0 is an attracting periodic point. For any of the other $3^n - 1$ periodic points, say z_m, one will have that $f^{(k)}(z_m) = z_j$, where z_j also is a periodic point. Furthermore, $|z_m| = 1$ for all these points. Therefore $(f^{(n)})'(z_m) = 3^n$ and these periodic points are all repelling.

Problem 7. Show that all complex quadratic polynomials are topologically conjugate to a polynomial of the form $q_c(z) = z^2 + c$, where $c \in \mathbb{C}$.

Solution 7. Let

$$f(z) = az^2 + bz + d, \quad a, b, d \in \mathbb{C}$$

and assume that

$$h(z) = \alpha z + \beta, \quad \alpha, \beta \in \mathbb{C}$$

is a topological conjugacy between f and q_c, i.e. $h(f(z)) = q_c(h(z))$. Then follows that

$$\alpha f(z) + \beta = h^2(z) + c$$

i.e.

$$\alpha a z^2 + \alpha b z + \alpha d + \beta = \alpha^2 z^2 + 2\alpha\beta z + \beta^2 + c.$$

Comparing the coefficients of the powers of z on either side of the equation, one finds the equations

$$\alpha a = \alpha^2, \quad \alpha b = 2\alpha\beta, \quad \alpha d + \beta = \beta^2 + c,$$

which are solved to yield the following constants

$$\alpha = a \quad \beta = \frac{1}{2} b \quad c = ad + \frac{1}{2} b(1 - b).$$

Obviously, α, β and c are uniquely determined for a given f.

Problem 8. Let $f : \mathbb{C} \to \mathbb{C}$ be a differentiable complex function. Let z^* be a fixed point of f. If $|f'(z^*)| < 1$, then the stable set of z^* contains a neighbourhood of z^*. If $|f'(z^*)| > 1$, then there is a neighbourhood of z^* all of whose points must leave the neighbourhood under iteration of f. Apply this statement to the function

$$f(z) = 2\sin(z) + z.$$

Solution 8. The fixed points are the solutions of the equation

$$2\sin(z^*) + z^* = z^*.$$

Thus $\sin(z^*) = 0$ and the fixed points are given by $z^* = n\pi$, where $n \in \mathbb{Z}$. Now

$$\frac{df(z)}{dz} = 2\cos(z) + 1.$$

Thus

$$\frac{df(z^* = n\pi)}{dz} = 2\cos(n\pi) + 1.$$

If n is even we have $df(n\pi)/dz = 3 > 0$. If n is odd we have $df(n\pi)/dz = -1 < 0$.

Problem 9. Let $\lambda = \exp(2\pi i\gamma)$ and $\gamma = \frac{1}{2}(\sqrt{5} - 1)$. Study the complex map

$$f(z) = \frac{\lambda}{2}z^2 + 1 - \frac{\lambda}{2}.$$

Solution 9. The map has a fixed point at $z = 1$, with linearization pure rotation by $2\pi\gamma$. Since $df/dz = \lambda z$ the map has a critical point, z_0, at $z = 0$. The iterates of the critical point

$$\{\, z_0,\ z_1 = f(z_0),\ z_2 = f(f(z_0)), \dots \}$$

lie on a fractal curve (the boundary of the Siegel disc).

Problem 10. Let $z \in \mathbb{C}$. Consider the map

$$f_{\mu,\gamma}(z) = \frac{\mu z(1 - \gamma z)}{1 + \mu(1 - \gamma)z}, \qquad \mu > 1,\ 0 \le \gamma \le 1.$$

The map $f_{\mu,\gamma}$ has a fixed point at $1 - 1/\mu$. Consider the invertible map

$$\phi(z) = z + 1 - \frac{1}{\mu}.$$

Let

$$F_{\mu,\gamma}(z) := (\phi^{-1} \circ f_{\mu,\gamma} \circ \phi)(z).$$

What happens to the fixed points under the map?

Solution 10. Obviously 0 is a fixed point of $F_{\mu,\gamma}$.

Problem 11. Consider the functions $f(z) = z^3$ and $h(z) = z + 1/z$. Find a function p such that

$$h(f(z)) = p(h(z)). \quad (1)$$

Solution 11. From (1) we obtain

$$z^3 + 1/z^3 = p(z + 1/z).$$

Letting $w = z + 1/z$ and solving for z gives

$$z_+ = \frac{1}{2}(w + \sqrt{w^2 - 4}), \qquad z_- = \frac{1}{2}(w - \sqrt{w^2 - 4}).$$

Since $z_- = 1/z_+$ we obtain

$$\begin{aligned} p(w) &= \left(\frac{1}{2}(w + \sqrt{w^2 - 4})\right)^3 + \left(\frac{1}{2}(w - \sqrt{w^2 - 4})\right)^3 \\ &= \frac{1}{8}(2w^3 + 6w(w^2 - 4)) \\ &= w^3 - 3w. \end{aligned}$$

Thus $p(z) = z^3 - 3z$.

Problem 12. Let $P(z)$ be a polynomial of degree $n \geq 2$ with distinct zeros $\zeta_1, \ldots, \zeta_n$. Show that

$$\sum_{j=1}^{n} \frac{1}{P'(\zeta_j)} = 0$$

where $'$ denotes the derivative, i.e. $P'(\zeta) \equiv dP(z = \zeta)/dz$.

Solution 12. When $n = 2$, the result is a straightforward calculation. For $n > 2$, we use induction. Let

$$P(z) = (z - \zeta_n)Q(z)$$

where the roots of $Q(z)$ are $\zeta_1, \ldots, \zeta_{n-1}$. Suppose all of the ζ_j are distinct. By *partial fraction* we have

$$\frac{1}{Q(z)} = \sum_{j=1}^{n} \frac{1}{(Q'(\zeta_j))(z - \zeta_j)}.$$

Consequently

$$\frac{1}{Q(\zeta_n)} = \sum_{j=1}^{n-1} \frac{1}{Q'(\zeta_j)(\zeta_n - \zeta_j)}.$$

Since

$$P'(z) = Q(z) + zQ'(z) - \zeta_n Q'(z)$$

we have $P'(\zeta_n) = Q(\zeta_n)$ and $P'(\zeta_j) = (\zeta_j - \zeta_n)Q'(\zeta_j)$ for $j < n$. Hence

$$\sum_{j=1}^{n} \frac{1}{P'(\zeta_j)} = \frac{1}{Q(\zeta_n)} + \sum_{j=1}^{n-1} \frac{1}{(\zeta_j - \zeta_n)Q'(\zeta_j)} = \frac{1}{Q(\zeta_n)} - \frac{1}{Q(\zeta_n)} = 0.$$

Problem 13. Show the following. Let $P(z)$ be a polynomial. Then either

1. $P(z)$ has a fixed point q with $P'(q) = 1$
2. $P(z)$ has a fixed point q with $|P'(q)| > 1$.

Solution 13. Define

$$R(z) := P(z) - z.$$

Then the roots of R are the fixed points of P. If the roots of R are not all distinct, then there exists ζ with $R(\zeta) = 0$ and $R'(\zeta) = 0$. Then $P(\zeta) = \zeta$ and $P'(\zeta) = 1$. Hence we may assume that the roots of R are all distinct. Let $\zeta_1, \ldots, \zeta_n$ be these roots. We have

$$\sum_{i=1}^{n} \frac{1}{P'(\zeta_i) - 1} = \sum_{i=1}^{n} \frac{1}{R'(\zeta_i)} = 0. \tag{1}$$

Suppose $|P'(\zeta_i)| \leq 1$ but $P'(\zeta_i) \neq 1$ for all i. Then $P'(\zeta_i) - 1$ lies in the circle $|z + 1| \leq 1$ minus the origin. Therefore $1/(P'(\zeta_i) - 1)$ is well defined and lies in the left-half plane. However since

$$\sum_{i=1}^{n} \frac{1}{P'(\zeta_i) - 1} = 0 \tag{2}$$

at least one of the $P'(\zeta_i) - 1$ must lie in the region $\Re(z) \geq 0$. This contradiction establishes the result.

Problem 14. Given any four complex numbers z_1, z_2, z_3, z_4 one defines the *cross ratio* $[z_1, z_2, z_3, z_4]$ by

$$[z_1, z_2, z_3, z_4] := \frac{(z_1 - z_2)(z_3 - z_4)}{(z_1 - z_3)(z_2 - z_4)}.$$

Show that

$$[z_1, z_2, z_3, z_4] = [f(z_1), f(z_2), f(z_3), f(z_4)]$$

for the *Möbius transformation*

$$f(z) = \frac{az+b}{cz+d}$$

with $ad - bc \neq 0$.

Solution 14. This is a typical problem to solve with computer algebra. A Maxima program is

```
f1: (a*z1+b)/(c*z1+d);
f2: (a*z2+b)/(c*z2+d);
f3: (a*z3+b)/(c*z3+d);
f4: (a*z4+b)/(c*z4+d);
ex1: ((f1-f2)*(f3-f4))/((f1-f3)*(f2-f4));
ex2: ratsimp(ex1);
ex3: ((z1-z2)*(z3-z4))/((z1-z3)*(z2-z4));
R: ex2-ex3;
R: ratsimp(R);
```

Problem 15. The *Schwarzian derivative* of a C^3 function f of one complex variable is defined by

$$(Sf)(z) := \left(\frac{f''(z)}{f'(z)}\right)' - \frac{1}{2}\left(\frac{f''(z)}{f'(z)}\right)^2 = \frac{f'''(z)}{f'(z)} - \frac{3}{2}\left(\frac{f''(z)}{f'(z)}\right)^2.$$

The Schwarzian derivative can also be written as

$$(Sf)(y) := 6\lim_{x\to y}\left(\frac{f'(x)f'(y)}{(f(x)-f(y))^2} - \frac{1}{(x-y)^2}\right).$$

Find the Schwarzian derivative of

$$g(z) = \frac{az+b}{cz+d}, \qquad ad - bc \neq 0.$$

Solution 15. We obtain $(Sg)(z) = 0$. If f and g have negative Schwarzian derivative, then $f \circ g$ has negative Schwarzian derivative.

Problem 16. The properties of the logistic map $x_{t+1} = 4x_t(1-x_t)$ ($x_0 \in [0,1], t = 0,1,2,\ldots$) are well-known. Let $z = x_1 + ix_2$, where $x_1, x_2 \in \mathbb{R}$. Study the map

$$z_{t+1} = 4z_t(1-z_t).$$

With $z_t = x_{1,t} + ix_{2,t}$ we can write

$$x_{1,t+1} = 4(x_{1,t} - x_{1,t}^2 + x_{2,t}^2), \qquad x_{2,t+1} = 4x_{2,t}(1 - 2x_{1,t}).$$

Thus we have the map $\mathbf{f} : \mathbb{R}^2 \to \mathbb{R}^2$

$$f_1(x_1, x_2) = 4(x_1 - x_1^2 + x_2^2), \qquad f_2(x_1, x_2) = 4x_2(1 - 2x_1)$$

with

$$\frac{\partial f_1}{\partial x_1} = 4 - 8x_1, \quad \frac{\partial f_1}{\partial x_2} = 8x_2, \quad \frac{\partial f_2}{\partial x_1} = -8x_2, \quad \frac{\partial f_2}{\partial x_2} = 4(1 - 2x_1).$$

With $z = re^{i\phi}$ $(r \geq 0)$ we could also write

$$r_{t+1}e^{\phi_{t+1}} = 4r_t e^{i\phi_t}(1 - r_t e^{i\phi_t}).$$

With $\phi_t = 0$ for all t we end up at the logistic map. First find the fixed points and study their stability.

Solution 16. The fixed points are determined by the equations

$$4(x_1 - x_1^2 + x_2^2) = x_1, \qquad 4x_2(1 - 2x_1) = x_2.$$

The solutions provides the fixed points

$$(0, 0), \quad (3/4, 0), \quad (3/8, 3i/8), \quad (3/8, -3i/8).$$

Only the first two are real. The variational equation is

$$y_{1,t+1} = (4 - 8x_{1,t})y_{1,t} + 8x_{2,t}y_{2,t}$$
$$y_{2,t+1} = -8x_{2,t}y_{1,t} + 4(1 - 2x_{1,t})y_{2,t}.$$

Inserting the fixed point $(0, 0)$ provides

$$y_{1,t+1} = 4y_{1,t}, \quad y_{2,t+1} = 4y_{2,t}.$$

Thus the fixed point $(0, 0)$ is unstable. Inserting the fixed point $(3/4, 0)$ provides

$$y_{1,t+1} = 2y_{1,t}, \quad y_{2,t+1} = -2y_{2,t}$$

and the fixed point $(3/4, 0)$ is also unstable.

Problem 17. Find the solution of the map $f : \mathbb{C} \to \mathbb{C}$, $f(z) = 4z(1 - z)$ or written as difference equation

$$z_{t+1} = 4z_t(1 - z_t), \qquad t = 0, 1, \ldots$$

Solution 17. We obtain

$$z_t = \sin^2(2^t \arcsin(\sqrt{z_0})).$$

So if $z_0 = 0$, then $z_t = 0$ since $\arcsin(0) = 0$. If $z_0 = 1$ we have $\arcsin(1) = \pi/2$ and $z_t = \sin^2(2^t\pi/2)$. Note that $\arcsin(-1) = \pi/2$.

Problem 18. Consider the smooth *vector field* in $\mathbb{C}^2$

$$V = (z_1 - z_2)^2 \frac{\partial}{\partial z_1} + (z_1 - z_2)^2 \frac{\partial}{\partial z_2} \equiv (z_1 - z_2)^2 \left(\frac{\partial}{\partial z_1} + \frac{\partial}{\partial z_2} \right).$$

Find the map $\mathbf{f} : \mathbb{R}^2 \to \mathbb{R}^2$ generated by

$$f_1(z_1, z_2) = z_1' = e^V z_1, \qquad f_2(z_1, z_2) = z_2' = e^V z_2.$$

Find a fixed point of this map.

Solution 18. Since $Vz_1 = (z_1 - z_2)^2$, $Vz_2 = (z_1 - z_2)^2$ and

$$V(Vz_1) = 0, \qquad V(Vz_2) = 0$$

we obtain

$$f_1(z_1, z_2) = z_1 + (z_1 - z_2)^2, \qquad f_2(z_1, z_2) = z_2 + (z_1 - z_2)^2.$$

A fixed point is given by $(z_1^*, z_2^*) = (0, 0)$.

Problem 19. Consider the complex map

$$z_{t+1} = z_t - \frac{z_t^3 - 1}{3z_t^2}, \qquad t = 0, 1, 2, \ldots$$

with $z_0 \neq 0$.
(i) Find the fixed points.
(ii) Consider the initial value

$$z_0 = \frac{1}{2} + i\frac{\sqrt{3}}{2} \equiv e^{i\pi/3}.$$

Find z_1.

Solution 19. This is *Newton's method* to solve $z^3 - 1 = 0$. Thus the fixed points are

$$1, \quad e^{2\pi i/3}, \quad e^{4\pi i/3}.$$

(ii) We obtain

$$z_1 = e^{-i2\pi/3} \equiv e^{i4\pi/3}$$

which is a fixed point.

Problem 20. Schwarz lemma. Let $\mathbb{D} := \{z : |z| < 1\}$ be the open unit disk in the complex plane $\mathbb{C}$ centered at the origin. Let $f : \mathbb{D} \to \mathbb{D}$ be a holomorphic map with $f(0) = 0$ (i.e. 0 is a fixed point). Then

$$|f(z)| \leq |z| \text{ for all } z \text{ in } \mathbb{D} \text{ and } |df(z=0)/dz| \leq 1.$$

Apply the lemma to

$$f(z) = \frac{1}{2}(\sin(z) + z).$$

Solution 20. A fixed point is $z^* = 0$. We have

$$\frac{df(z)}{dz} = \frac{1}{2}\cos(z) + \frac{1}{2}$$

and thus

$$\frac{df(z^* = 0)}{dz} = 1.$$

Problem 21. Let $\mathbb{D} := \{z : |z| < 1\}$. Let $f : \mathbb{D} \to \mathbb{D}$ be analytic and assume that the map f is not an elliptic Möbius transformation nor the identity. Then there is an $\tilde{z} \in \bar{\mathbb{D}}$ such that $f^{(n)}(z) \to \tilde{z}$ for all $z \in \mathbb{D}$. Consider the map

$$f(z) = \frac{z + 1/2}{1 + z/2}.$$

(i) Find the fixed points of $f : \bar{\mathbb{D}} \to \bar{\mathbb{D}}$.
(ii) Find $f(0)$, $f(f(0))$, $f(f(f(0)))$, ...
(iii) Find $f(-1/2)$, $f(f(-1/2))$, $f(f(f(-1/2)))$, ...

Solution 21. (i) From the equation $f(z^*) = z^*$ we obtain the fixed points $+1$ and -1.
(ii) We find

$$f(0) = \frac{1}{2}, \quad f(1/2) = \frac{4}{5}, \quad f(4/5) = \frac{13}{14}.$$

The sequence tends to the fixed point $+1$.
(iii) We obtain

$$f(-1/2) = 0, \quad f(0) = \frac{1}{2}, \quad f(1/2) = \frac{4}{5}.$$

As in (i) the sequence tends to the fixed point +1.

Problem 22. Let $c \neq 0$. Consider the transcendental equation

$$f_c(z) = c - z\tan(z) = 0.$$

(i) Find the *Newton map*

$$N_{f_c}(z) := z - \frac{f_c(z)}{df_c(z)/dz}.$$

(ii) Show that if $c = z\tan(z)$ we have

$$N_{f_c}(z) = z.$$

Solution 22. (i) We have

$$\frac{df_c(z)}{dz} = \frac{-\sin(z)\cos(z) - z}{\cos^2(z)}.$$

Thus

$$N_{f_c}(z) = \frac{z^2 + c\cos^2(z)}{z + \sin(z)\cos(z)}.$$

(iii) We have

$$N_{f_c}(z)|_{c=z\tan(z)} = z.$$

Problem 23. Consider the function $f : \mathbb{C} \to \mathbb{C}$

$$f(z) = z^4 + 1.$$

(i) Find roots of f, i.e. solve $f(z) = 0$.
(ii) Find the fixed points of f, i.e. solve $f(z^*) = z^*$.
(iii) Find the Newton map

$$N_f(z) := z - \frac{f(z)}{df(z)/dz}.$$

Solution 23. (i) Utilizing that $e^{i\pi} = -1$ we find the four roots

$$\frac{1}{\sqrt{2}}(1+i), \quad -\frac{1}{\sqrt{2}}(1+i), \quad \frac{1}{\sqrt{2}}(1-i), \quad \frac{1}{\sqrt{2}}(-1+i).$$

(ii) One can show that there is no real fixed point. We solve numerically $f(z^*) = z^*$ i.e.

$$1 - z^* + (z^*)^4 = 0$$

with the R command

```
polyroot(c(1.0,-1.0,0.0,0.0,1.0))
```

which provides r_1, $\bar{r}_1$, r_2, $\bar{r}_2$ with

```
r1 = 0.7271361 + 0.4300143i
r2 = -0.7271361 + 0.9340993i
```

(iii) With $df/dz = 4z^3$ we obtain the *Newton map*

$$N_f(z) = \frac{3z^4 - 1}{4z^3}.$$

For example we have $N_f(i) = i/2$.

Problem 24. Let $a \in \mathbb{R}$ and $a > 0$. Consider the analytic function $f : \mathbb{C} \to \mathbb{C}$

$$f(w) = w^2 + 2aw$$

with the fixed points $w^* = 0$ and $w^* = 1 - 2a$. Consider the initial value problem for the nonlinear differential equation in the complex plane

$$\frac{dw}{dz} = \frac{1}{2(w+a)}$$

with the corresponding vector field

$$V = \frac{1}{2(w+a)}\frac{d}{dw}.$$

(i) Find $V(f(w))$ and $V(V(f(w)))$.
(ii) Find the solution of the initial value problem via the *Lie series*

$$w(z) = e^{zV} w\big|_{w=w(0)}.$$

(iii) Find the solution of the initial value problem applying

$$f(w(z)) \equiv w^2(z) + 2aw(z) \equiv f\left(e^{zV}\big|_{w=w(0)}\right).$$

Solution 24. (i) We have

$$V(f(w)) = 1, \qquad V(V(f(w))) = 0.$$

(ii) Direct calculation yields

$$w(z) = e^{zV} w\big|_{w=w(0)} = w(0) + \sum_{k=1}^{\infty} \binom{1/2}{k} \frac{z^k}{(w(0)+a)^{2k-1}}.$$

(iii) Since

$$f\left(e^{zV}w\big|_{w=w(0)}\right) = e^{zV}f(w)\big|_{w=w(0)}$$
$$= (f(w)+z)\big|_{w=w(0)}$$
$$= (w^2+2aw)\big|_{w=w(0)} + z$$
$$= w^2(0)+2aw(0)+z$$

we obtain

$$w(z) = -a + \sqrt{(w^2(0)+a)^2+z}.$$

Problem 25. Let n be a positive integer. Then the solutions of $z^n = 1$ are called the roots of unity. For example, if $n = 4$ we have $z_1 = 1$, $z_2 = i$, $z_3 = -1$, $z_4 = -i$. Write a C++ program using the class `complex<double>` that stores the unit roots in an array for a given n.

Solution 25. The roots are given by $(i = \sqrt{-1})$

$$\exp(ij2\pi/n), \qquad j = 0, 1, \ldots, n-1.$$

Now we can write

$$\exp(ij2\pi/n) \equiv \cos(j2\pi/n) + i\sin(j2\pi/n).$$

```
// unitroots.cpp

#include <iostream>
#include <complex>
#include <cmath>
using namespace std;

void unitroots(complex<double>* a,int n)
{
 const double pi = 3.14159265358979323846;
 for(int j=0;j<n;j++)
  a[j] = complex<double>(cos(j*2.0*pi/n),sin(j*2.0*pi/n));
}

int main(void)
{
 int n = 4;
 complex<double>* a = new complex<double>[n];
 unitroots(a,n);
 for(int j=0;j<n;j++)
 {
```

```
cout << "real part a[" << j << "] = " << a[j].real() << endl;
cout << "imag part a[" << j << "] = " << a[j].imag() << endl;
}
delete[] a;
return 0;
}
```

2.3.2 Supplementary Problems

Problem 1. The *Ikeda laser map* is given by

$$z \mapsto r + c_2 z \exp\left(i\left(c_1 - \frac{c_3}{1+|z|^2}\right)\right)$$

where r, c_1, c_2, c_3 are real positive bifurcation parameters. With $z = x + iy$ and $x, y \in \mathbb{R}$ write the map as difference equation for x_t and y_t. Find the fixed points. Study the stability of the fixed points. Show that the map admits chaotic behaviour for certain parameter values.

Problem 2. In the study of the Potts model the following complex map appears

$$z_{t+1} = \frac{abz_t - 1/2}{az_t + b - 3/2}, \qquad t = 0, 1, \ldots$$

where $a > 0$ and $b > 0$. Find the fixed points and study their stability.

Problem 3. Consider the difference equation $z_{t+1} = 4z_t(1 - z_t)$ with $t = 0, 1, 2, \ldots$ and $z_0 \in \mathbb{C}$. Show that with $z_t = x_t + iy_t$ and x_t, $y_t \in \mathbb{R}$ we can write

$$\begin{aligned} x_{t+1} &= 4x_t(1 - x_t) + 4y_t^2 \\ 1 - x_{t+1} &= x_t^2 - 2x_t(1 - x_t) + (1 - x_t)^2 - 4y_t^2 \\ y_{t+1} &= 4y_t - 8x_t y_t \equiv 4(1 - x_t)y_t - 4x_t y_t \end{aligned}$$

or in matrix form

$$\begin{pmatrix} x_{t+1} \\ 1 - x_{t+1} \\ y_{t+1} \end{pmatrix} =$$

$$\begin{pmatrix} 0 & 2 & 0 & 2 & 0 & 0 & 0 & 0 & 4 \\ 1 & -1 & 0 & -1 & 1 & 0 & 0 & 0 & -4 \\ 0 & 0 & -2 & 0 & 0 & 2 & -2 & 2 & 0 \end{pmatrix} \left(\begin{pmatrix} x_t \\ 1 - x_t \\ y_t \end{pmatrix} \otimes \begin{pmatrix} x_t \\ 1 - x_t \\ y_t \end{pmatrix} \right).$$

Problem 4. Let $z \in \mathbb{C}$ and consider the analytic map

$$f(z) = \exp(z).$$

Find the fixed points of f, i.e. solve the equation $z = f(z)$. Setting $z = x + iy$ $(x, y \in \mathbb{R})$ we have

$$x + iy = \exp(x + iy) \equiv e^x e^{iy} = e^x(\cos(y) + i\sin(y)).$$

Thus we have to solve the system of nonlinear equations

$$e^x \cos(y) - x = 0, \qquad e^x \sin(y) - y = 0.$$

Obviously we have to solve this system numerically.

Problem 5. Is the function $f : \mathbb{C} \to \mathbb{C}$

$$f(z) = \frac{e^z - 1}{e^z + 1}$$

analytic? Find the fixed points.

Problem 6. Consider $f : \mathbb{C} \to \mathbb{C}$

$$f(z) = z + 2z^2 + 3z^3.$$

(i) Find the fixed points of f. Note that $z^* = 0$ is one of the fixed points.
(ii) Find the fixed points of df/dz.
(iii) Find the fixed points of d^2f/dz^2.

Problem 7. Let $\mathbb{R}^+$ be the nonnegative real numbers. Consider $f : \mathbb{R}^+ \to \mathbb{R}$, $f(x) = \sqrt{x}$. Then $x^* = 0$ and $x^* = 1$ are fixed points of the map, i.e solutions of $\sqrt{x^*} = x^*$. The fixed point $x^* = 0$ is unstable and the fixed point $x^* = 1$ is stable. Find the fixed points of the map $f : \mathbb{C} \to \mathbb{C}$, $f(z) = \sqrt{z}$. Set $z = re^{i\phi}$ with $r \geq 0$ and $\phi \in [0, 2\pi)$. Study the stability of the fixed points. Iterate the map f, i.e. find $f(f(z))$ and the fixed points of $f(f(z))$.

2.4 Higher Dimensional Maps

2.4.1 Solved Problems

Problem 1. Consider the three-dimensional analytic map $\mathbf{f} : \mathbb{R}^3 \to \mathbb{R}^3$

$$f_1(x_1, x_2, x_3) = x_1x_2 - x_3, \quad f_2(x_1, x_2, x_3) = x_1, \quad f_3(x_1, x_2, x_3) = x_2$$

or written as a system of difference equations

$$x_{1,t+1} = x_{1,t}x_{2,t} - x_{3,t}, \quad x_{2,t+1} = x_{1,t}, \quad x_{3,t+1} = x_{2,t} \qquad t = 0, 1, 2, \ldots$$

(i) Consider the (volume) three differential form $\omega = dx_1 \wedge dx_2 \wedge dx_3$. Find $\mathbf{f}^*(\omega)$.
(ii) Study the initial values $x_{1,0} = x_{3,0} = 1/2$, $x_{2,0} = -1$.
(iii) Find the fixed points of $\mathbf{f}$ and study their stability.
(iv) Find the fixed points of the second iterate.
(v) Find the inverse of the map.

Solution 1. (i) With

$$df_1 = x_2dx_1 + x_1dx_2 - dx_3, \quad df_2 = dx_1, \quad df_3 = dx_2$$

we obtain

$$\mathbf{f}^*(\omega) = -dx_1 \wedge dx_2 \wedge dx_3.$$

Thus the Jacobian determinant is equal to -1 and the map is invertible.
(ii) We find

$$x_{1,1} = -1, \; x_{2,1} = \frac{1}{2}, \; x_{3,1} = -1, \quad x_{1,2} = \frac{1}{2}, \; x_{2,2} = -1, \; x_{3,2} = \frac{1}{2}.$$

Thus we have a periodic orbit with period 2.
(iii) From the fixed point equations we obtain two fixed points $(0, 0, 0)$ and $(2, 2, 2)$. The variational equation is

$$\begin{pmatrix} y_{1,t+1} \\ y_{2,t+1} \\ y_{3,t+1} \end{pmatrix} = \begin{pmatrix} x_{2,t} & x_{1,t} & -1 \\ 1 & 0 & 0 \\ 0 & 1 & 0 \end{pmatrix} \begin{pmatrix} y_{1,t} \\ y_{2,t} \\ y_{3,t} \end{pmatrix}.$$

For the fixed point $(0, 0, 0)$ the eigenvalues of the matrix on the right-hand side are given by -1, $e^{i\pi/3} \equiv \frac{1}{2}(1 + i\sqrt{2})$, $e^{-i\pi/3} \equiv \frac{1}{2}(1 - i\sqrt{2})$. The eigenvalues for the fixed point $(2, 2, 2)$ the eigenvalues of the matrix on the right-hand side are given by -1, $\frac{1}{2}(3 + \sqrt{5})$, $\frac{1}{2}(3 - \sqrt{5})$. Note that the last two eigenvalues are positive.
(iv) The second iterate of $\mathbf{f}$ is given by

$$\begin{aligned} f_1(f_1, f_2, f_3) &= (x_1x_2 - x_3)x_1 - x_2 \\ f_2(f_1, f_2, f_3) &= x_1x_2 - x_3 \\ f_3(f_1, f_2, f_3) &= x_1. \end{aligned}$$

The fixed points are $(0, 0, 0)$ and with $\gamma \in \mathbb{R}$ and $\gamma \neq 1$ we have

$$(x_1^* = \gamma, x_2^* = \gamma/(\gamma - 1), x_3^* = \gamma).$$

With $\gamma = 2$ we obtain the fixed point $(2, 2, 2)$ of $\mathbf{f}$.
(v) Note that

$$\begin{pmatrix} \partial f_1/\partial x_1 & \partial f_1/\partial x_2 & \partial f_1/\partial x_3 \\ \partial f_2/\partial x_1 & \partial f_2/\partial x_2 & \partial f_2/\partial x_3 \\ \partial f_3/\partial x_1 & \partial f_3/\partial x_2 & \partial f_3/\partial x_3 \end{pmatrix} = \begin{pmatrix} x_2 & x_1 & -1 \\ 1 & 0 & 0 \\ 0 & 1 & 0 \end{pmatrix}.$$

The determinant of this 3×3 matrix is -1 and thus independent of x_1, x_2. The inverse map is given

$$f_1^{-1}(x_1, x_2, x_3) = x_2, \quad f_2^{-1}(x_1, x_2, x_3) = x_3, \quad f_3^{-1}(x_1, x_2, x_3) = x_2 x_3 - x_1.$$

Problem 2. Consider the analytic map $\mathbf{f} : \mathbb{R}^3 \to \mathbb{R}^3$

$$f_1(x_1, x_2, x_3) = x_1 x_2 - x_3, \quad f_2(x_1, x_2, x_3) = x_1, \quad f_3(x_1, x_2, x_3) = x_2$$

and the *metric tensor field* in $\mathbb{R}^3$

$$g = dx_1 \otimes dx_1 + dx_2 \otimes dx_2 + dx_3 \otimes dx_3$$

Find $\mathbf{f}^*(g)$.

Solution 2. With

$$df_1 = x_2 dx_1 + x_1 dx_2 - dx_3, \quad df_2 = dx_1, \quad df_3 = dx_2$$

and

$$g = df_1 \otimes df_1 + df_2 \otimes df_2 + df_3 \otimes df_3$$

we obtain

$$\begin{aligned} \mathbf{f}^*(g) = & (1 + x_2^2) dx_1 \otimes dx_1 + (1 + x_1^2) dx_2 \otimes dx + dx_3 \otimes dx_3 \\ & + x_1 x_2 dx_1 \otimes dx_2 + x_1 x_2 dx_2 \otimes dx_1 \\ & - x_1 dx_2 \otimes dx_3 - x_1 dx_3 \otimes dx_2 \\ & - x_2 dx_1 \otimes dx_3 - x_2 dx_3 \otimes dx_1 \end{aligned}$$

with the corresponding symmetric matrix

$$\begin{pmatrix} 1 + x_2^2 & x_1 x_2 & -x_2 \\ x_1 x_2 & 1 + x_1^2 & -x_1 \\ -x_2 & -x_1 & 1 \end{pmatrix}.$$

The determinant of this matrix is equal to 1. What are the eigenvalues?

Problem 3. Consider the three-dimensional analytic map $\mathbf{f} : \mathbb{R}^3 \to \mathbb{R}^3$

$$f_1(x_1, x_2, x_3) = x_2 x_3, \quad f_2(x_1, x_2, x_3) = x_1 x_3, \quad f_3(x_1, x_2, x_3) = x_1 x_2$$

or written as a system difference equations

$$x_{1,t+1} = x_{2,t}x_{3,t}, \quad x_{2,t+1} = x_{1,t}x_{3,t}, \quad x_{3,t+1} = x_{1,t}x_{2,t} \quad t = 0, 1, 2, \ldots$$

(i) Let $\omega = dx_1 \wedge dx_2 \wedge dx_3$. Find $\mathbf{f}^*(\omega)$.
(ii) Find the orbit for $x_{1,0} = 1/2$, $x_{2,0} = 1$, $x_{3,0} = 3/2$.
(iii) Find the fixed points and study their stability.
(iv) Find the second iterate $\mathbf{f}^{(2)}$ and their fixed points.

Solution 3. (i) With

$$df_1 = x_2 dx_3 + x_3 dx_2, \quad df_2 = x_1 dx_3 + x_3 dx_1, \quad df_3 = x_1 dx_2 + x_2 dx_1$$

with obtain

$$\mathbf{f}^*(\omega) = 2x_1x_2x_3\,\omega.$$

Thus the map is not invertible.
(ii) We obtain

$$\begin{aligned} x_{1,1} &= 3/2, & x_{1,2} &= 3/8, & x_{1,3} &= 9/32 \\ x_{2,1} &= 3/4, & x_{2,2} &= 3/4, & x_{2,3} &= 27/64 \\ x_{3,1} &= 1/2, & x_{3,2} &= 9/8, & x_{3,3} &= 9/32. \end{aligned}$$

Since $|x_{1,3}| < 1$, $|x_{2,3}| < 1$, $|x_{3,3}| < 1$ and the map only contains the products x_1x_2, x_2x_3, x_1x_3 the orbit tends to the fixed point $(0, 0, 0)$.
(iii) The fixed points are

$$(0,0,0), \quad (1,1,1), \quad (1,-1,-1), \quad (-1,1,-1), \quad (-1,-1,1).$$

The variational equation is

$$\begin{pmatrix} y_{1,t+1} \\ y_{2,t+1} \\ y_{3,t+1} \end{pmatrix} = \begin{pmatrix} 0 & x_{3,t} & x_{2,t} \\ x_{3,t} & 0 & x_{1,t} \\ x_{2,t} & x_{1,t} & 0 \end{pmatrix} \begin{pmatrix} y_{1,t} \\ y_{2,t} \\ y_{3,t} \end{pmatrix}.$$

For the fixed point $(0,0,0)$ the eigenvalues of the matrix are 0, 0, 0. The fixed point is stable. For the fixed point $(1,1,1)$ the eigenvalues are 2 $(1\times)$ and -1 $(2\times)$. Also the other three fixed point admit these eigenvalues. Thus these fixed points are unstable.
(iv) The second iterate $\mathbf{f}^{(2)}$ is

$$f_1(f_1, f_2, f_3) = x_1^2 x_2 x_3, \quad f_2(f_1, f_2, f_3) = x_1 x_2^2 x_3, \quad f_3(f_1, f_2, f_3) = x_1 x_2 x_3^2.$$

The fixed points are $(0,0,0)$ and with $\gamma_1, \gamma_2 \in \mathbb{R}$ and $\gamma_1 \neq 0$, $\gamma_2 \neq 0$

$$(x_1^* = 1/(\gamma_1\gamma_2), x_2^* = \gamma_1, x_3^* = \gamma_2).$$

Obviously these fixed points of $\mathbf{f}^{(2)}$ include the fixed points of $\mathbf{f}$ (as it must be). For example for $\gamma_1 = \gamma_2 = 1$ we obtain $(1,1,1)$.

Problem 4. Consider the system of difference equations

$$\begin{aligned} x_{1,t+1} &= x_{1,t}^2 + x_{2,t}^2 x_{3,t} \\ x_{2,t+1} &= x_{1,t}x_{2,t} + x_{1,t}x_{2,t}x_{3,t} \\ x_{3,t+1} &= x_{1,t}^2 x_{3,t} + x_{3,t}^2 \end{aligned}$$

where $t = 0, 1, 2, \ldots$ and $x_{1,0}$, $x_{2,0}$, $x_{3,0}$ are the initial conditions. Find the fixed points and study their stability.

Solution 4. Besides the trivial fixed point $P_0 = (0,0,0)$ we have the fixed points $P_1 = (0,0,1)$, $P_2 = (\gamma,\gamma,\gamma)$ $(\gamma = (\sqrt{5}-1)/2)$, $P_3 = (1, x_2, 0)$ where x_2 is arbitrary. The line of fixed points P_3 has two marginal directions $\lambda_{1,2} = 1$ and one repulsive direction $\lambda_3 = 2$ independent of x_2. For the fixed point P_1 we have $\lambda_{1,2} = 0$ and $\lambda_3 = 2$. For the fixed point P_2 we have $\lambda_1 = 2\gamma - 1 < 1$, $\lambda_2 = 2\gamma > 1$ and $\lambda_3 = 2 + \gamma^2 > 1$.

Problem 5. Let $f_j : \mathbb{R}^3 \to \mathbb{R}$ be analytic functions. Consider the autonomous system of first order ordinary differential equations

$$\frac{dx_1}{dt} = f_1(x_1,x_2,x_3), \quad \frac{dx_2}{dt} = f_2(x_1,x_2,x_3), \quad \frac{dx_3}{dt} = f_3(x_1,x_2,x_3)$$

with the corresponding smooth *vector field*

$$V = f_1(x_1,x_2,x_3)\frac{\partial}{\partial x_1} + f_2(x_1,x_2,x_3)\frac{\partial}{\partial x_2} + f_3(x_1,x_2,x_3)\frac{\partial}{\partial x_3}.$$

From the *Lie series expansion* we have

$$\begin{aligned} e^{tV}x_1 &= x_1 + tVx_1 + \cdots = x_1 + tf_1(\mathbf{x}) + \cdots \\ e^{tV}x_2 &= x_2 + tVx_2 + \cdots = x_2 + tf_2(\mathbf{x}) + \cdots \\ e^{tV}x_3 &= x_3 + tVx_3 + \cdots = x_3 + tf_3(\mathbf{x}) + \cdots \end{aligned}$$

With $t = 1$ we consider the analytic maps

$$g_1(\mathbf{x}) = x_1 + f_1(\mathbf{x}), \quad g_2(\mathbf{x}) = x_2 + f_2(\mathbf{x}), \quad g_3(\mathbf{x}) = x_3 + f_3(\mathbf{x}).$$

Assume that $\mathbf{x}^* = (x_1, x_2, x_3)$ is a fixed point of the system of first order differential equations, i.e.

$$f_1(\mathbf{x}^*) = 0, \quad f_2(\mathbf{x}^*) = 0, \quad f_3(\mathbf{x}^*) = 0.$$

Solution 5. Obviously we have

$$x_1^* = x_1^* + f_1(\mathbf{x}^*), \quad x_2^* = x_2^* + f_2(\mathbf{x}^*), \quad x_3^* = x_3^* + f_3(\mathbf{x}^*).$$

Problem 6. Consider the function $\mathbf{f} : \mathbb{R}^3 \to \mathbb{R}^3$ with $f_1 : \mathbb{R}^3 \to \mathbb{R}$, $f_2 : \mathbb{R}^3 \to \mathbb{R}$, $f_3 : \mathbb{R}^3 \to \mathbb{R}$ are analytic functions. Written as system of difference equations we have

$$x_{1,t+1} = f_1(x_{1,t}, x_{2,t}, x_{3,t})$$
$$x_{2,t+1} = f_2(x_{1,t}, x_{2,t}, x_{3,t})$$
$$x_{3,t+1} = f_3(x_{1,t}, x_{2,t}, x_{3,t}).$$

Then the variational equation is given by

$$y_{1,t+1} = \frac{\partial f_1}{\partial x_1} y_{1,t} + \frac{\partial f_1}{\partial x_2} y_{2,t} + \frac{\partial f_1}{\partial x_3} y_{3,t}$$
$$y_{2,t+1} = \frac{\partial f_2}{\partial x_1} y_{1,t} + \frac{\partial f_2}{\partial x_2} y_{2,t} + \frac{\partial f_2}{\partial x_3} y_{3,t}$$
$$y_{3,t+1} = \frac{\partial f_3}{\partial x_1} y_{1,t} + \frac{\partial f_3}{\partial x_2} y_{2,t} + \frac{\partial f_3}{\partial x_3} y_{3,t}.$$

Assume that $\mathbf{u} =$ and $\mathbf{v}$ also satisfy the variational equation.
(i) Let $\mathbf{e}_1$, $\mathbf{e}_2$, $\mathbf{e}_3$ be an orthonormal basis in $\mathbb{R}^3$ and

$$\mathbf{y}_t = y_{1,t}\mathbf{e}_1 + y_{2,t}\mathbf{e}_2 + y_{3,t}\mathbf{e}_3, \qquad \mathbf{u}_t = u_{1,t}\mathbf{e}_1 + u_{2,t}\mathbf{e}_2 + u_{3,t}\mathbf{e}_3.$$

Let $\wedge$ be the *exterior product* (also called Grassmann product or wedge product). Then

$$\mathbf{y}_t \wedge \mathbf{u}_t = (y_{1,t}u_{2,t} - y_{2,t}u_{1,t})\mathbf{e}_1 \wedge \mathbf{e}_2$$
$$+(y_{2,t}u_{3,t} - y_{3,t}u_{2,t})\mathbf{e}_2 \wedge \mathbf{e}_3$$
$$+(y_{3,t}u_{1,t} - y_{1,t}u_{3,t})\mathbf{e}_3 \wedge \mathbf{e}_1$$

where we utilized that $\mathbf{e}_j \wedge \mathbf{e}_j = 0$ and $\mathbf{e}_j \wedge \mathbf{e}_k = -\mathbf{e}_k \wedge \mathbf{e}_j$ for $k \neq j$. Find the time-evolutions of

$$w_{12,t} = y_{1,t}u_{2,t} - y_{2,t}u_{1,t}, \quad w_{23,t} = y_{2,t}u_{3,t} - y_{3,t}u_{2,t}, \quad w_{31,t} = y_{3,t}u_{1,t} - y_{1,t}u_{3,t}$$

i.e. find $w_{12,t+1}$, $w_{23,t+1}$, $w_{31,t+1}$.
(ii) Let

$$\mathbf{y}_t = y_{1,t}\mathbf{e}_1 + y_{2,t}\mathbf{e}_2 + y_{3,t}\mathbf{e}_3$$
$$\mathbf{u}_t = u_{1,t}\mathbf{e}_1 + u_{2,t}\mathbf{e}_2 + u_{3,t}\mathbf{e}_3$$
$$\mathbf{v}_t = v_{1,t}\mathbf{e}_1 + v_{2,t}\mathbf{e}_2 + v_{3,t}\mathbf{e}_3$$

with $\mathbf{y}_t$, $\mathbf{u}_t$, $\mathbf{v}_t$ satisfying the variational equation. Then

$$\mathbf{y} \wedge \mathbf{u} \wedge \mathbf{v} = (y_{1,t}u2, tv_{3,t} - y_{2,t}u_{1,t}v_{3,t} + y_{3,t}u_{1,t}v_{2,t} - y_{1,t}u_{3,t}v_{2,t} + y_{2,t}u_{3,t}v_{1,t} - y_{3,t}u_{2,t}v_{1,t})\, \mathbf{e}_1 \wedge \mathbf{e}_2 \wedge \mathbf{e}_3.$$

Obviously the coefficient of $\mathbf{e}_1 \wedge \mathbf{e}_2 \wedge \mathbf{e}_3$ is the determinant of the 3×3 matrix

$$\begin{pmatrix} y_{1,t} & y_{2,t} & y_{3,t} \\ u_{1,t} & u_{2,t} & u_{3,t} \\ v_{1,t} & v_{2,t} & v_{3,t} \end{pmatrix}.$$

Find the time evolution of

$$w_{123,t} = \det \begin{pmatrix} y_{1,t} & y_{2,t} & y_{3,t} \\ u_{1,t} & u_{2,t} & u_{3,t} \\ v_{1,t} & v_{2,t} & v_{3,t} \end{pmatrix}$$

i.e. find $w_{123,t+1}$.

Solution 6. Straightforward calculation yields

$$\begin{aligned} w_{12,t+1} = & \left(\frac{\partial f_1}{\partial x_1} \frac{\partial f_2}{\partial x_2} - \frac{\partial f_2}{\partial x_1} \frac{\partial f_1}{\partial x_2} \right) w_{12,t} \\ & + \left(\frac{\partial f_1}{\partial x_2} \frac{\partial f_2}{\partial x_3} - \frac{\partial f_2}{\partial x_2} \frac{\partial f_1}{\partial x_3} \right) w_{23,t} \\ & + \left(\frac{\partial f_1}{\partial x_3} \frac{\partial f_2}{\partial x_1} - \frac{\partial f_1}{\partial x_1} \frac{\partial f_2}{\partial x_3} \right) w_{31,t} \end{aligned}$$

$$\begin{aligned} w_{23,t+1} = & \left(\frac{\partial f_2}{\partial x_1} \frac{\partial f_3}{\partial x_2} - \frac{\partial f_3}{\partial x_1} \frac{\partial f_2}{\partial x_2} \right) w_{12,t} \\ & + \left(\frac{\partial f_2}{\partial x_2} \frac{\partial f_3}{\partial x_3} - \frac{\partial f_3}{\partial x_2} \frac{\partial f_2}{\partial x_3} \right) w_{23,t} \\ & + \left(\frac{\partial f_2}{\partial x_3} \frac{\partial f_3}{\partial x_1} - \frac{\partial f_3}{\partial x_3} \frac{\partial f_2}{\partial x_1} \right) w_{31,t} \end{aligned}$$

$$\begin{aligned} w_{31,t+1} = & \left(\frac{\partial f_3}{\partial x_1} \frac{\partial f_1}{\partial x_2} - \frac{\partial f_3}{\partial x_2} \frac{\partial f_1}{p x_1} \right) w_{12,t} \\ & + \left(\frac{\partial f_3}{\partial x_2} \frac{\partial f_1}{\partial x_3} - \frac{\partial f_3}{\partial x_3} \frac{\partial f_1}{\partial x_2} \right) w_{23,t} \\ & + \left(\frac{\partial f_3}{\partial x_3} \frac{\partial f_1}{\partial x_1} - \frac{\partial f_3}{\partial x_1} \frac{\partial f_1}{\partial x_3} \right) w_{31,t}. \end{aligned}$$

(ii) Let

$$F = \begin{pmatrix} \partial f_1/\partial x_1 & \partial f_1/\partial x_2 & \partial f_1/\partial x_3 \\ \partial f_2/\partial x_1 & \partial f_2/\partial x_2 & \partial f_2/\partial x_2 \\ \partial f_3/\partial x_1 & \partial f_3/\partial x_2 & \partial f_3/\partial x_3 \end{pmatrix}.$$

Then

$$w_{123,t+1} = \det(F) w_{123,t}.$$

Problem 7. (i) Calculate the *weight matrix* W given by

$$W = \mathbf{x}_0 \mathbf{x}_0^T + \mathbf{x}_1 \mathbf{x}_1^T - 2I_8$$

for the *Hopfield network* which stores the two patterns

$$\mathbf{x}_0 = (1,1,1,1,1,1,1,1)^T, \qquad \mathbf{x}_1 = (1,1,-1,-1,-1,-1,1,1)^T$$

(ii) The iteration is given by

$$\mathbf{s}(t+1) = \text{sign}(W\mathbf{s}(t))$$

with $t = 0,1,2,\ldots$. Which of these two vectors are fixed points under iteration of the network?
(iii) Consider the initial vector

$$\mathbf{s}(t=0) = (-1,1,1,1,1,1,1,-1)^T.$$

Calculate the evolution of this vector under synchronous evolution. Does $\mathbf{s}(t)$ approach a fixed point?

Solution 7. (i) We obtain the symmetric 8×8 matrix

$$W = \begin{pmatrix} 0 & 2 & 0 & 0 & 0 & 0 & 2 & 2 \\ 2 & 0 & 0 & 0 & 0 & 0 & 2 & 2 \\ 0 & 0 & 0 & 2 & 2 & 2 & 0 & 0 \\ 0 & 0 & 2 & 0 & 2 & 2 & 0 & 0 \\ 0 & 0 & 2 & 2 & 0 & 2 & 0 & 0 \\ 0 & 0 & 2 & 2 & 2 & 0 & 0 & 0 \\ 2 & 2 & 0 & 0 & 0 & 0 & 0 & 2 \\ 2 & 2 & 0 & 0 & 0 & 0 & 2 & 0 \end{pmatrix}.$$

(ii) We find

$$\text{sgn}(W\mathbf{x}_0) = \mathbf{x}_0, \qquad \text{sgn}(W\mathbf{x}_1) = \mathbf{x}_1$$

and thus $\mathbf{x}_0$ and $\mathbf{x}_1$ are fixed points of the map.
(iii) We have

$$\mathbf{s}(1) = \text{sgn}(W\mathbf{s}(0)) = (1,-1,1,1,1,1,-1,1)^T$$

and

$$\mathbf{s}(2) = \text{sgn}(W\mathbf{s}(1)) = \mathbf{s}(0).$$

So we have an oscillating behaviour and $\mathbf{s}(0)$ does not tend to a fixed point.

Problem 8. Let $t = 0, 1, 2, \ldots$, $K > 0$ and $N > 1$. Consider the map

$$p_j(t+1) = p_j(t) + \frac{K}{2\pi}(\sin(2\pi(x_{j+1}(t) - x_j(t))) - \sin(2\pi(x_j(t) - x_{j-1}(t))))$$
$$x_j(t+1) = x_j(t) + p_j(t+1)$$

where $j = 1, 2, \ldots, N$ and $x_{j+N} \equiv x_j$, $p_{j+N} \equiv p_j$ (periodic boundary conditions). Find

$$\sum_{j=1}^{N} dx_j(t+1) \wedge dp_j(t+1)$$

where $\wedge$ denotes the *exterior product* (also called wedge or Grassmann product).

Solution 8. Since

$$dx_j(t) \wedge dx_j(t) = 0, \qquad dp_j(t+1) \wedge dp_j(t+1) = 0$$

and $dx_j(t+1) = dx_j(t) + dp_j(t+1)$ and utilizing the period boundary conditions we obtain

$$\sum_{j=1}^{N} dx_j(t+1) \wedge dp_j(t+1) = \sum_{j=1}^{N} dx_j(t) \wedge dp_j(t).$$

Problem 9. The *Denman-Beavers iteration* for the square root of an $n \times n$ matrix A with no eigenvalues on $\mathbb{R}^-$ is

$$Y_{k+1} = \frac{1}{2}(Y_k + Z_k^{-1})$$
$$Z_{k+1} = \frac{1}{2}(Z_k + Y_k^{-1})$$

with $k = 0, 1, 2, \ldots$ and $Z_0 = I_n$ and $Y_0 = A$. The iteration has the properties that

$$\lim_{k\to\infty} Y_k = A^{1/2}, \qquad \lim_{k\to\infty} Z_k = A^{-1/2}$$

and, for all k,

$$Y_k = AZ_k, \quad Y_kZ_k = Z_kY_k, \quad Y_{k+1} = \frac{1}{2}(Y_k + AY_k^{-1}).$$

(i) Can the Denman-Beavers iteration be applied to the matrix

$$A = \begin{pmatrix} 1 & 1 \\ 1 & 2 \end{pmatrix} ?$$

(ii) Find Y_1 and Z_1.

Solution 9. (i) For the eigenvalues we find

$$\lambda_{1,2} = \frac{1}{2}(3 \pm \sqrt{5}).$$

Thus there is no eigenvalue on $\mathbb{R}^-$ and the Denman-Beavers iteration can be applied.
(ii) We have $\det(A) = 1$. For the inverse of A we find

$$A^{-1} = \begin{pmatrix} 2 & -1 \\ -1 & 1 \end{pmatrix}$$

and therefore

$$Y_1 = \begin{pmatrix} 1 & 1/2 \\ 1/2 & 3/2 \end{pmatrix}, \quad Z_1 = \begin{pmatrix} 3/2 & -1/2 \\ -1/2 & 1/2 \end{pmatrix}.$$

For $n \to \infty$, Y_n converges to

$$\begin{pmatrix} 0.89443 & 0.44721 \\ 0.44721 & 1.34164 \end{pmatrix}.$$

Problem 10. The number $\pi/2$ can be calculated using the iteration

$$x_{k+1} = x_k y_k, \qquad y_{k+1} = \sqrt{2y_k/(y_k+1)}, \quad k = 0, 1, 2, \ldots$$

where $x_0 = 1$, $y_0 = \sqrt{2}$. Then

$$\lim_{k \to \infty} x_k = \frac{\pi}{2}.$$

Write a C++ program that implements this iteration and thus finds an approximation of $\pi/2$.

Solution 10. Using a `do-while` we have

```
// pihalf.cpp

#include <iostream>
#include <cmath>
```

```
using namespace std;

int main(void)
{
 double x0 = 1.0; double y0 = sqrt(2.0);
 double x1, y1, t;
 double eps = 0.0001;
 do
 {
 x1 = x0*y0; y1 = sqrt(2.0*y0/(y0+1.0));
 t = fabs(x0-x1);
 x0 = x1; y0 = y1;
 } while(t > eps);
 cout << "x1 = " << x1;
 return 0;
}
```

The output is `x1 = 1.57079`.

2.4.2 Supplementary Problems

Problem 1. Study the three-dimensional analytic map

$$f_1(x_1, x_2, x_3) = \sinh(x_1)\sinh(x_2) - \sinh(x_3)$$
$$f_2(x_1, x_2, x_3) = \sinh(x_1)$$
$$f_3(x_1, x_2, x_3) = \sinh(x_2).$$

Show that $(0, 0, 0)$ is a fixed point. Are there other fixed points?

Problem 2. Let $\epsilon \in (0, 1)$ and $f : \mathbb{R} \to \mathbb{R}$, $f(x) = 1 - 2x^2$. Study the globally coupled map

$$x_{0,t+1} = (1-\epsilon)f(x_{0,t}) + \frac{\epsilon}{3}\sum_{j=0}^{2} f(x_{j,t})$$

$$x_{1,t+1} = (1-\epsilon)f(x_{1,t}) + \frac{\epsilon}{3}\sum_{j=0}^{2} f(x_{j,t})$$

$$x_{2,t+1} = (1-\epsilon)f(x_{2,t}) + \frac{\epsilon}{3}\sum_{j=0}^{2} f(x_{j,t})$$

with the initial values $x_{0,0}$, $x_{1,0}$, $x_{2,0}$.

Problem 3. Consider the three-dimensional map

$$x_{1,t+1} = r_1 - x_{2,t}^2 - r_2 x_{3,t}, \quad x_{2,t+1} = x_{1,t}, \quad x_{3,t+1} = x_{2,t}$$

where $t = 0, 1, 2, \ldots$ and $r_1 > 0$, $r_2 > 0$ are the bifurcation parameters. Show that the map can show hyperchaotic behaviour (depending on r_1 and r_2) i.e. two one-dimensional Liapunov exponents can be positive.

Problem 4. Consider the three-dimensional map $\mathbf{f} : \mathbb{R}^3 \to \mathbb{R}^3$

$$f_1(x, y, z) = xz \left(\frac{(xz + 2y^2)(1 + xy^2 + zy^2)^2}{(x^2z^2 + y^2z^2 + x^2y^2)^2(1 + 2x^2y^2z^2)} \right)^{1/3}$$

$$f_2(x, y, z) = y \left(\frac{(x^2y^2 + y^2z^2 + z^2x^2)(xz + 2y^2)}{(1 + xy^2 + zy^2)(1 + 2x^2y^2z^2)} \right)^{1/3}$$

$$f_3(x, y, z) = \left(\frac{(x^2y^2 + y^2z^2 + z^2x^2)(1 + xy^2 + zy^2)^2}{(xz + 2y^2)^2(1 + 2x^2y^2z^2)} \right)^{1/3}$$

which appears at renormalization group transformation. Show that the fixed points are given by (i) $x^* = z^* = 1$, y arbitrary, (ii) $x^* = y^* = z^* = 0$, (iii) $x^* = 0$, $y^* = 1$, $z^* = 1$.

Problem 5. Let $k = 0, 1, 2, \ldots$ and

$$y_k = \int_0^1 \frac{x^k}{1 + x + x^2} dx$$

with

$$y_0 = \int_0^1 \frac{1}{1 + x + x^2} dx = \frac{2}{\sqrt{3}} \arctan \left(\frac{2x + 1}{\sqrt{3}} \right),$$

$$y_1 = \int_0^1 \frac{x}{1 + x + x^2} dx = \frac{1}{2} \ln(1 + x + x^2) - \frac{1}{\sqrt{3}} \arctan(\left(\frac{2x + 1}{\sqrt{3}} \right).$$

Show that

$$y_{k+2} + y_{k+1} + y_k = \frac{1}{k+1}, \quad k = 0, 1, 2, \ldots$$

Problem 6. Solve the initial value problem for the third order difference equation

$$x_{t+2} = x_t(x_{t+1} - x_{t-1})$$

with $t = 1, 2, \ldots$ and $x_0 = 1$, $x_1 = 1/2$, $x_2 = 1$.

Problem 7. Consider the *Fibonacci trace map*

$$x_{t+1} = 2x_t x_{t-1} - x_{t-2} \tag{1}$$

where $t = 2, 3, \ldots$. Map (1) is a discrete dynamical system with various physical applications. Show that the Fibonacci trace map is reversible and possesses the invariant

$$\tilde{I}(x_{t-1}, x_t, x_{t+1}) = x_{t-1}^2 + x_t^2 + x_{t+1}^2 - 2x_{t-1}x_n x_{n+1} - 1. \qquad (2)$$

If $I = 0$ and $|x_i| \leq 1$, $i = 1, 2, 3$, we are in a region of homogeneous chaos. This follows from a semi-conjugacy to a hyperbolic automorphism of the torus which makes the system pseudo-Anosov

$$x_t = \cos(2\pi\theta_t), \qquad \theta_{t+1} = \theta_t + \theta_{t-1}. \qquad (3)$$

Problem 8. The *Arnold cat map* is given by

$$\begin{pmatrix} x_{1,t+1} \\ x_{2,t+1} \end{pmatrix} = \begin{pmatrix} 1 & 1 \\ 1 & 2 \end{pmatrix} \begin{pmatrix} x_{1,t} \\ x_{2,t} \end{pmatrix} \mod 1.$$

The map shows chaotic behaviour. Let $a, b \in \mathbb{N}$, i.e. a, b are positive integers. Consider the 2×2 matrix

$$M(a,b) = \begin{pmatrix} 1 & a \\ b & ab+1 \end{pmatrix}$$

with $\det(M(a,b)) = 1$. It contains the matrix given above with $a = b = 1$.
(i) Show that the matrix is invertible. Find the inverse matrix.
(ii) What are the conditions on $a, b \in \mathbb{N}$ such that $M(a,b)$ is a normal matrix?
(iii) Find the eigenvalues and normalized eigenvectors of $M(a,b)$.
(iv) Find the two one-dimensional Liapunov exponents for the map ($t = 0, 1, 2, \ldots$)

$$\begin{pmatrix} x_{1,t+1} \\ x_{2,t+1} \end{pmatrix} = \begin{pmatrix} 1 & a \\ b & ab+1 \end{pmatrix} \begin{pmatrix} x_{1,t} \\ x_{2,t} \end{pmatrix} \mod 1.$$

(v) Let $\otimes$ be the Kronecker product. Calculate $M(a,b) \otimes M(c,d)$ with $c, d \in \mathbb{N}$. Find the eigenvalues and normalized eigenvectors of $M(a,b) \otimes M(c,d)$. Utilize the results from (iii).
(vi) Find the four one-dimensional Liapunov exponents for the map ($t = 0, 1, 2, \ldots$)

$$\begin{pmatrix} x_{1,t+1} \\ x_{2,t+1} \\ x_{3,t+1} \\ x_{4,t+1} \end{pmatrix} = (M(a,b) \otimes M(c,d)) \begin{pmatrix} x_{1,t} \\ x_{2,t} \\ x_{3,t} \\ x_{4,t} \end{pmatrix}.$$

(vii) The *star product* of the matrices $M(a,b)$ and $M(c,d)$ is defined as

$$M(a,b) \star M(c,d) = \begin{pmatrix} 1 & 0 & 0 & c \\ 0 & 1 & a & 0 \\ 0 & b & ab+1 & 0 \\ d & 0 & 0 & cd+1 \end{pmatrix}.$$

Find the eigenvalues and normalized eigenvectors of $M(a,b) \star M(c,d)$. Utilize the result from (iii).
(viii) Find the four one-dimensional Liapunov exponents for the map

$$\begin{pmatrix} x_{1,t+1} \\ x_{2,t+1} \\ x_{3,t+1} \\ x_{4,t+1} \end{pmatrix} = (M(a,b) \star M(c,d)) \begin{pmatrix} x_{1,t} \\ x_{2,t} \\ x_{3,t} \\ x_{4,t} \end{pmatrix} \mod 1.$$

Problem 9. Consider the matrices

$$A_0 = \begin{pmatrix} 0 & 1 \\ -1 & 0 \end{pmatrix}, \qquad B_0 = \begin{pmatrix} 0 & i \\ i & 0 \end{pmatrix}.$$

Let $n = 0, 1, 2, \ldots$. Study the sequence of matrices

$$A_{n+1} = A_n B_n, \qquad B_{n+1} = A_n.$$

Discuss. Is these sequence of matrices periodic?

Problem 10. Consider the alphabet $\Sigma = \{U, V, W\}$, axiom: $\omega = U$ and the set of production rules

$$U \mapsto UVW, \quad V \mapsto UV, \quad W \mapsto U.$$

(i) Apply it to $U = \sigma_1$, $V = \sigma_2$, $W = \sigma_3$ and matrix multiplication. Is the sequence periodic?
(ii) Apply it to $U = \sigma_1$, $V = \sigma_2$, $W = \sigma_3$ and the Kronecker product.

Problem 11. Let $a > 0$, $b > 0$. Study the three dimensional map $\mathbf{f} : \mathbb{R}^3 \to \mathbb{R}^3$

$$f_1(x_1, x_2, x_3) = a - x_2^2 - bx_3, \quad f_2(x_1, x_2, x_3) = x_1, \quad f_3(x_1, x_2, x_3) = x_2$$

or written as difference equations

$$x_{1,t+1} = a - x_{2,t}^2 - bx_{3,t}, \quad x_{2,t+1} = x_{1,t}, \quad x_{3,t+1} = x_{2,t}$$

where $t = 0, 1, \ldots$. Show that $dx_1 \wedge dx_2 \wedge dx_3$ is invariant under the map. Are the differential one forms

$$x_1 dx_2 + x_2 dx_3 + x_3 dx_1, \qquad x_1 dx_2 - x_2 dx_3 + x_3 dx_1$$

invariant under the map $\mathbf{f}$?

Problem 12. Study the three-dimensional map $\mathbf{f} : \mathbb{R}^3 \to \mathbb{R}^3$

$$f_1(x_1, x_2, x_3) = x_2^2 - x_3^2, \quad f_2(x_1, x_2, x_3) = x_3^2 - x_1^2, \quad f_3(x_1, x_2, x_3) = x_1^2 - x_2^2.$$

Note that $(x_1^*, x_2^*, x_3^*) = (0,0,0)$ is a fixed point.

Problem 13. Study the map $\mathbf{f} : \mathbb{Z} \times \mathbb{Z} \times \mathbb{Z} \to \mathbb{Z} \times \mathbb{Z} \times \mathbb{Z}$

$$f_1(x_1, x_2, x_3) = x_1 - x_2 x_3$$
$$f_2(x_1, x_2, x_3) = -x_2 + x_1 x_3$$
$$f_3(x_1, x_2, x_3) = x_3 - x_1 x_2.$$

Show that the fixed points are

$$(0,0,0), \quad (1,0,0), \quad (0,0,1), \quad (-1,0,0), \quad (0,0,-1).$$

Show that $(x_1 = 1, x_2 = 2, x_3 = 1)$ provides an eventually periodic orbit. Is the map invertible?

2.5 Bitwise Maps

2.5.1 Solved Problems

Problem 1. Let $x_{1,t}, x_{2,t} \in \{0,1\}$ and $t = 0,1,2,\ldots$. We denote by $\oplus$ the XOR-operation, by $+$ the OR-operation and by $\cdot$ the AND-operation. Consider the map

$$x_{1,t+1} = x_{1,t} + x_{2,t}, \qquad x_{2,t+1} = x_{1,t} \oplus x_{2,t}$$

with $t = 0,1,2,\ldots$.
(i) Find the fixed points of the map, i.e. solve the set of equations

$$x_1 + x_2 = x_1, \qquad x_1 \oplus x_2 = x_2.$$

(ii) Let $x_{1,0} = x_{2,0} = 1$. Does $(x_{1,t}, x_{2,t})$ tend to a fixed point for $t \to \infty$?

Solution 1. (i) The systems of equations

$$x_1 + x_2 = x_1, \qquad x_1 \oplus x_2 = x_2$$

admits the solution $x_1 = 0$, $x_2 = 0$.
(ii) With $x_{1,0} = 1$, $x_{2,0} = 1$ we obtain

$$x_{1,1} = 1, \quad x_{2,1} = 0, \qquad x_{1,2} = 1, \quad x_{2,2} = 1.$$

Thus we have a period orbit.

Problem 2. The *Feynman gate* is a 2 input/2 output gate given by

$$f_1(x_1, x_2) = x_1' = x_1$$
$$f_2(x_1, x_2) = x_2' = x_1 \oplus x_2$$

The fixed points are $(0,0)$ and $(0,1)$.
(i) Give the truth table for the Feynman gate.
(ii) Show that copying can be implemented using the Feynman gate.
(iii) Show that the complement can be implemented using the Feynman gate.
(iv) Is the Feynman gate invertible?

Solution 2. (i) The truth table is

x_1	x_2	x_1'	x_2'
0	0	0	0
0	1	0	1
1	0	1	1
1	1	1	0

(ii) Setting $x_2 = 0$, we have $x_2' = x_1 \oplus 0 = x_1$. Thus we have a copy.
(iii) Setting $x_2 = 1$, we have $x_2' = x_1 \oplus 1 = \bar{x}_1$. Thus we generated the complement.
(iv) From the truth table we see that the transformation is invertible. The inverse transformation can be found as follows. Since $x_1 \oplus x_1 = 0$ we have

$$x_1' \oplus x_2' = x_1 \oplus x_1 \oplus x_2 = 0 \oplus x_2 = x_2.$$

Thus $x_1 = x_1'$, $x_2 = x_1' \oplus x_2'$.

Problem 3. Consider the boolean function

$$f_1(x_1, x_2) = x_1' = x_2, \quad f_2(x_1, x_2) = x_2' = x_1 \oplus x_2.$$

Give the truth table. Discuss. The only fixed point is $(0,0)$.

Solution 3. The truth table is

x_1	x_2	x_1'	x_2'
0	0	0	0
0	1	1	1
1	0	0	1
1	1	1	0

Thus the boolean function is invertible.

Problem 4. Can the boolean function $\mathbf{f} : \{0,1\}^2 \to \{0,1\}^2$

$$f_1(x_1, x_2) = (x_1 \cdot x_2) + (x_1 \cdot x_2), \qquad f_2(x_1, x_2) = (x_1 + x_2) \cdot (x_1 + x_2)$$

be simplified? Is the map invertible?

Solution 4. Yes. We have $f_1(x_1, x_2) = x_1 \cdot x_2$, $f_2(x_1, x_2) = x_1 + x_2$. No, the map is not invertible.

Problem 5. Can the boolean function $\mathbf{f} : \{0,1\}^2 \to \{0,1\}^2$

$$f_1(x_1, x_2) = (x_1 \cdot x_2) \oplus (x_1 \cdot x_2), \qquad f_2(x_1, x_2) = (x_1 \oplus x_2) \cdot (x_1 \oplus x_2)$$

be simplified? Is the map invertible?

Solution 5. Yes. We have

$$f_1(x_1, x_2) = 0, \qquad f_2(x_1, x_2) = x_1 \oplus x_2.$$

Owing to f_1 the map is not invertible.

Problem 6. Let n be a positive integer. Then 2^n bitstrings of length n can be formed.
(i) How many boolean maps for $\mathbf{f} : \{0,1\}^2 \to \{0,1\}^2$ exist?
(ii) How many invertible boolean maps for $\mathbf{f} : \{0,1\}^2 \to \{0,1\}^2$ exist?

Solution 6. (i) There are $4 \cdot 4 \cdot 4 \cdot 4 = 256$ such maps exist.
(ii) There are $4! = 24$ such maps exist.

Problem 7. Consider the 3-input/3-output gate given by

$$\begin{aligned} f_1(x_1, x_2, x_3) &= x_1' = x_1 \\ f_2(x_1, x_2, x_3) &= x_2' = x_1 \oplus x_2 \\ f_3(x_1, x_2, x_3) &= x_3' = x_1 \oplus x_2 \oplus x_3. \end{aligned}$$

(i) Give the truth table.
(ii) Is the transformation invertible.

Solution 7. (i) The truth table is given by

x_1	x_2	x_3	x_1'	x_2'	x_3'
0	0	0	0	0	0
0	0	1	0	0	1
0	1	0	0	1	1
0	1	1	0	1	0
1	0	0	1	1	1
1	0	1	1	1	0
1	1	0	1	0	0
1	1	1	1	0	1

(ii) From the truth table we see that the transformation is invertible, i.e. we have a $1-1$ map. The inverse transformation is given by

$$x_1 = x_1'$$
$$x_2 = x_1' \oplus x_2'$$
$$x_3 = x_1' \oplus x_2' \oplus x_3'.$$

Problem 8. Let $x_1, x_2, x_3 \in \{0,1\}$. Consider the maps

$$f_1(x_1,x_2,x_3) = (x_1 \cdot (x_2 + x_3))$$
$$f_2(x_1,x_2,x_3) = (x_1 \cdot (x_2 \oplus x_3))$$
$$f_3(x_1,x_2,x_3) = (x_1 + (x_2 \oplus x_3)).$$

Let $x_1 = x_2 = x_3 = 1$. Find

$$f_1(1,1,1),\quad f_2(1,1,1),\quad f_3(1,1,1)$$

and

$$f_1(f_1(1,1,1)),\quad f_2(f_2(1,1,1)),\quad f_3(f_3(1,1,1)).$$

Solution 8. We have

$$f_1(1,1,1) = 1, \quad f_2(1,1,1) = 0, \quad f_3(1,1,1) = 1$$

and

$$f_1(1,0,1) = 1, \quad f_1(1,0,1) = 1, \quad f_3(1,0,1) = 1.$$

Thus we have a periodic orbit. The fixed point of the map is $(0,0,0)$.

Problem 9. Let $t = 0,1,2,\ldots$ and $x_t, y_t, z_t \in \{0,1\}$ and the map

$$x_{1,t+1} = x_{2,t} \cdot x_{3,t}, \quad x_{2,t+1} = x_{3,t} + x_{1,t}, \quad x_{3,t+1} = x_{1,t} \oplus x_{2,t}$$

where $\cdot$ denotes the AND operation, $+$ the OR operation and $\oplus$ the XOR operation.
(i) Find the fixed points of the map.
(ii) Solve the map with the initial condition $x_{1,0} = 1$, $x_{2,0} = 1$, $x_{3,0} = 1$. Does the solution tend to a fixed point?

Solution 9. (i) The system of equations

$$x_2 \cdot x_3 = x_1, \quad x_3 + x_1 = x_2, \quad x_1 \oplus x_2 = x_3$$

provides the only solution $x_1 = 0$, $x_2 = 0$, $x_3 = 0$.
(ii) With the initial condition $x_{1,0} = 1$, $x_{2,0} = 1$, $x_{3,0} = 1$ we obtain

$$x_{1,1} = 1, \quad x_{2,1} = 1, \quad x_{3,1} = 0$$

$$x_{1,2} = 0, \quad x_{2,2} = 1, \quad x_{3,2} = 0$$

$$x_{1,3} = 0, \quad x_{2,3} = 0, \quad x_{3,3} = 1$$

$$x_{1,4} = 0, \quad x_{2,4} = 1, \quad x_{3,4} = 0$$

$$x_{1,5} = 0, \quad x_{2,5} = 0, \quad x_{3,5} = 1.$$

Thus we have a periodic orbit.

2.5.2 Supplementary Problems

Problem 1. Let $x_0, y_0 \in \{0, 1\}$. Solve the boolean equations

$$x_{1,t+1} = x_{1,t} \oplus x_{2,t}, \qquad x_{2,t+1} = x_{1,t} \cdot x_{2,t}$$

where $x_{1,0} = 1$, $x_{2,0} = 1$ and $t = 0, 1, \ldots$. Find the fixed points. Does the sequence $x_{1,t}$, $x_{2,t}$ tend to a fixed point?

Problem 2. (i) Let $s_1(0), s_2(0), s_3(0) \in \{+1, -1\}$. Study the time-evolution ($t = 0, 1, 2, \ldots$) of the coupled system of equations

$$s_1(t+1) = s_2(t)s_3(t)$$
$$s_2(t+1) = s_1(t)s_3(t)$$
$$s_3(t+1) = s_1(t)s_2(t)$$

for the eight possible initial conditions, i.e. (i) $s_1(0) = s_2(0) = s_3(0) = 1$, (ii) $s_1(0) = 1$, $s_2(0) = 1$, $s_3(0) = -1$, (iii) $s_1(0) = 1$, $s_2(0) = -1$, $s_3(0) = 1$, (iv) $s_1(0) = -1$, $s_2(0) = 1$, $s_3(0) = 1$, (v) $s_1(0) = 1$, $s_2(0) = -1$, $s_3(0) =$

-1, (vi) $s_1(0) = -1$, $s_2(0) = 1$, $s_3(0) = -1$, (vii) $s_1(0) = -1$, $s_2(0) = -1$, $s_3(0) = 1$, (viii) $s_1(0) = -1$, $s_2(0) = -1$, $s_3(0) = -1$. Which of these initial conditions are fixed points?

(ii) Let $s_1(0), s_2(0), s_3(0) \in \{+1, -1\}$. Study the time-evolution ($t = 01, 2, \ldots$) of the coupled system of equations

$$s_1(t+1) = s_2(t)s_3(t)$$
$$s_2(t+1) = s_1(t)s_2(t)s_3(t)$$
$$s_3(t+1) = s_1(t)s_2(t)$$

for the eight possible initial conditions, i.e. (i) $s_1(0) = s_2(0) = s_3(0) = 1$, (ii) $s_1(0) = 1$, $s_2(0) = 1$, $s_3(0) = -1$, (iii) $s_1(0) = 1$, $s_2(0) = -1$, $s_3(0) = 1$, (iv) $s_1(0) = -1$, $s_2(0) = 1$, $s_3(0) = 1$, (v) $s_1(0) = 1$, $s_2(0) = -1$, $s_3(0) = -1$, (vi) $s_1(0) = -1$, $s_2(0) = 1$, $s_3(0) = -1$, (vii) $s_1(0) = -1$, $s_2(0) = -1$, $s_3(0) = 1$, (viii) $s_1(0) = -1$, $s_2(0) = -1$, $s_3(0) = -1$. Which of these initial conditions are fixed points?

Problem 3. Let $x_1(0), x_2(0), x_3(0) \in \{0, 1\}$ and let $\oplus$ be the XOR-operation. Study the time-evolution ($t = 0, 1, 2, \ldots$) of the coupled system of equations

$$x_1(t+1) = x_2(t) \oplus x_3(t)$$
$$x_2(t+1) = x_1(t) \oplus x_3(t)$$
$$x_3(t+1) = x_1(t) \oplus x_2(t)$$

for the eight possible initial conditions, i.e. (i) $x_1(0) = x_2(0) = x_3(0) = 0$, (ii) $x_1(0) = 0$, $x_2(0) = 0$, $x_3(0) = 1$, (iii) $x_1(0) = 0$, $x_2(0) = 1$, $x_3(0) = 0$, (iv) $x_1(0) = 1$, $x_2(0) = 0$, $x_3(0) = 0$, (v) $x_1(0) = 0$, $x_2(0) = 1$, $x_3(0) = 1$, (vi) $x_1(0) = 1$, $x_2(0) = 0$, $x_3(0) = 1$, (vii) $x_1(0) = 1$, $x_2(0) = 1$, $x_3(0) = 0$, (viii) $x_1(0) = 1$, $x_2(0) = 1$, $x_3(0) = 1$. Which of these initial conditions are fixed points?

Problem 4. Let $\mathbb{Z}$ be the set of integers. Consider the two-dimensional lattice $\mathbb{Z} \times \mathbb{Z}$ and $(i, j) \in \mathbb{Z} \times \mathbb{Z}$. Let $t = 0, 1, 2, \ldots$. Consider the two-dimensional cellular automata ($s_{ij}(t) \in \{0, 1\}$)

$$s_{ij}(t+1) = s_{i,j+1}(t) \oplus s_{i-1,j}(t) \oplus s_{i,j}(t) \oplus s_{i+1,j}(t) \oplus s_{i,j-1}(t)$$

where $\oplus$ is the XOR-operation and at $t = 0$ we have $s_{0,0}(t = 0) = 1$ and 0 otherwise for all other lattice sites. Calculate $s_{i,j}(t = 1)$ and $s_{i,j}(t = 2)$. The four nearest neighbours around $(0, 0)$ are

$$(1, 0), \quad (0, 1), \quad (-1, 0), \quad (0, -1).$$

Problem 5. Given the truth table

x_1	x_2	x'_1	x'_2
0	0	0	1
0	1	1	0
1	0	1	1
1	1	0	0

Find f_1 and f_2.

Chapter 3

Fractals

3.1 Introduction

In this chapter we consider problems involving fractals. These include the Cantor set, the Mandelbrot set, the Julia set and iterated function systems. To investigate fractals one needs fractal dimensions. Fractal dimensions are the *capacity* (also called *box counting dimension*) and the Hausdorff dimension.

Definition. Let X be a subset of $\mathbb{R}^n$. Let $N(\epsilon)$ be the number of n-dimensional cubes (boxes) of side ϵ to cover the set X. The *capacity* D of X is defined as

$$D := \lim_{\epsilon \to 0} \frac{\ln N(\epsilon)}{\ln\left(\frac{1}{\epsilon}\right)}. \tag{1}$$

Definition. Let X be a subset of $\mathbb{R}^n$. A cover of X is a (possibly infinite) collection of balls the union of which contains X. The diameter of a cover $\mathcal{A}$ is the maximum diameter of the balls in $\mathcal{A}$. For $d, \epsilon > 0$, we define

$$\alpha(d, \epsilon) := \inf_{\substack{\mathcal{A} = \text{ coverof X} \\ \text{diam }(\mathcal{A}) \leq \epsilon}} \sum_{A \in \mathcal{A}} (\text{diam}(A))^d \tag{2}$$

and

$$\alpha(d) := \lim_{\epsilon \to 0} \alpha(d, \epsilon).$$

There is a unique d_0 such that

$$d < d_0 \Rightarrow \alpha(d) = \infty$$

$$d > d_0 \Rightarrow \alpha(d) = 0.$$

This d_0 is defined to be the *Hausdorff dimension* of X, written $HD(X)$.

In general we have the inequality

$$HD(X) \leq D(X).$$

Definition. A hyperbolic iterated function system consists of a complete metric space (X, d) together with a finite set of contraction mappings

$$f_n : X \to X$$

with respective contractivity factors s_n, for $n = 1, 2, \ldots, N$. The notation for the iterated function system is $\{X\,,\, f_n;\, n = 1, 2, \ldots, N\}$ and its contractivity factor is

$$s = \max\{s_n \;:\; n = 1, 2, \ldots, N\}.$$

The next theorem gives the properties of a hyperbolic iterated function system.

Theorem 1. Let $\{X\,,\, f_n \;:\; n = 1, 2, \ldots, N\}$ be a hyperbolic iterated function system with contractivity factor s. Then the transformation

$$W : \mathcal{H}(X) \to \mathcal{H}(X)$$

defined by

$$W(B) := \bigcup_{n=1}^{N} f_n(B)$$

for all $B \in \mathcal{H}(X)$, is a contraction mapping on the complete metric space $(\mathcal{H}(X), h(d))$ with contractivity factor s. That is

$$h(W(B), W(C)) \leq s \cdot h(B, C)$$

for all $B, C \in \mathcal{H}(X)$. Its unique fixed point, $A \in \mathcal{H}(X)$, obeys

$$A = W(A) = \bigcup_{n=1}^{N} f_n(A)$$

and is given by $A = \lim_{n\to\infty} W^{(n)}(B)$ for any $B \in \mathcal{H}(X)$. The fixed point $A \in \mathcal{H}(X)$ described in the theorem is called the attractor of the iterated function system.

Thus an iterated function system consisting of contractive similarity mappings has a unique attractor $A \subset \mathbb{R}^n$ which is invariant under the action of the system.

3.2 Solved Problems

Problem 1. The middle third *Cantor set* C (*ternary Cantor set*) is constructed as follows. We set

$$E_0 = [0,1]$$

$$E_1 = [0,1/3] \cup [2/3,1]$$

$$E_2 = [0,1/9] \cup [2/9,1/3] \cup [2/3,7/9] \cup [8/9,1]$$

$$\ldots = \ldots \tag{1}$$

In other words. Delete the middle third of the line segment $[0,1]$ and then the middle third from all the resulting segments and so on ad infinitum. The set defined by

$$C := \bigcap_{k=0}^{\infty} E_k \tag{2}$$

is called the standard Cantor set (or Cantor ternary set).
(i) Show that the Cantor set is of Lebesgue measure zero.
(ii) Show that there is a bijective mapping $f : C \mapsto [0,1]$.
(iii) Find the capacity of the interval $I = [0,2]$. Find the capacity of the standard Cantor set C.
(iv) Show that for the standard Cantor set $HD(C) \leq (\ln 2)/(\ln 3)$.

Solution 1. (i) We define

$$\chi_{E_k} := \begin{cases} 1 & x \in E_k \\ 0 & x \notin E_k. \end{cases}$$

Owing to the construction of the Cantor set the sets E_k are the union of 2^k pairwise disjoint intervals with length $1/3^k$. Thus

$$\int_0^1 \chi_{E_k} d\mu = 2^k \cdot \frac{1}{3^k} = \left(\frac{2}{3}\right)^k$$

and

$$\lim_{k\to\infty} \int_0^1 \chi_{E_k} d\mu = 0.$$

On the other hand $\chi_{E_{k+1}} \leq \chi_{E_k}$ and

$$\lim_{k\to\infty} \chi_{E_k} = \chi_C.$$

Therefore χ_C is integrable and

$$\mu(C) = \int_0^1 \chi_C d\mu = 0.$$

(ii) Every number $x \in [0,1]$ can be represented as (*binary representation*)

$$x = \sum_{j=1}^{\infty} a_j 2^{-j} \qquad a_j \in \{0,1\}.$$

Every $y \in C$ can be written in the form (*ternary representation*)

$$y = \sum_{j=1}^{\infty} c_j 3^{-j} \qquad c_j \in \{0,2\}.$$

Let $f : C \mapsto [0,1]$ be defined by

$$f(\sum_{j=1}^{\infty} c_j 3^{-j}) = \sum_{j=1}^{\infty} a_j 2^{-j}$$

with

$$a_j = \begin{cases} 0 \text{ for } c_j = 0 \\ 1 \text{ for } c_j = 2. \end{cases}$$

Obviously f is surjective. Since different numbers have different binary and ternary representations the mapping f is injective. Since the mapping f is surjective and injective we find that the mapping f is bijective.
(iii) Let $I = [0,2]$. Let $\epsilon > 0$ and $N \cdot \epsilon = 2$. Then $N(\epsilon) = 2/\epsilon$. Inserting this into the definition for the capacity we find

$$D = \lim_{\epsilon \to 0} \frac{\ln(2/\epsilon)}{\ln(1/\epsilon)} = 1.$$

To calculate the capacity of the Cantor set we note the following. In the first step of the construction we have $\epsilon = 1/3$ and $N = 2$. In the second step of the construction we have $\epsilon = 1/9$ and $N = 4$ and at the p-th step of the construction we find

$$\epsilon = \left(\frac{1}{3}\right)^p, \qquad N = 2^p.$$

Consequently, the capacity of the Cantor set is given by

$$D = \frac{\ln(2)}{\ln(3)}.$$

(iv) Owing to the construction of the standard Cantor set the following cover is obvious. For $n = 1, 2, \dots$, let $\mathcal{A}$ be the cover of C consisting of 2^n intervals of length $1/3^n$ each. We set

$$\sum_{A \in \mathcal{A}_n} (\text{diam}(A))^d = 1.$$

It follows that

$$2^n \left(\frac{1}{3^n}\right)^d = 1.$$

This leads to

$$d = \frac{\ln(2)}{\ln(3)}.$$

We conclude that with this d we have

$$\alpha\left(d, \frac{1}{3^n}\right) \leq \sum_{A \in \mathcal{A}_n} (\text{diam} A)^d = 1$$

for every n. Therefore

$$\alpha\left(\frac{\ln 2}{\ln 3}\right) \leq 1.$$

This implies that

$$HD(C) \leq \frac{\ln(2)}{\ln(3)}.$$

In fact one can show that $HD(C) = \ln(2)/\ln(3)$. Thus we have $D(C) = HD(C)$.

Problem 2. (i) Show that the middle third *Cantor set* can be generated from the two contractive maps $f_1 : [0,1] \to [0,1]$, $f_2 : [0,1] \to [0,1]$

$$f_1(x) = \frac{1}{3}x, \qquad f_2(x) = \frac{1}{3}x + \frac{2}{3}.$$

The fixed point of f_1 is $x^* = 0$ and the fixed point of f_2 is $x^* = 1$.
(ii) Calculate

$$f_1(1/2), \quad f_1(f_1(1/2)), \quad f_1(f_1(f_1(1/2)))$$

and

$$f_2(1/2), \quad f_2(f_2(1/2)), \quad f_2(f_2(f_2(1/2))).$$

Solution 2. (i) We have

$$f_1(0) = 0, \;\; f_1(1) = \frac{1}{3}, \qquad f_2(0) = \frac{2}{3}, \;\; f_2(1) = 1.$$

This provides the two remaining intervals $[0, 1/3]$ and $[2/3, 1]$ and the interval $(1/3, 2/3)$ is removed. Now

$$f_1(0) = 0, \;\; f_1(1/3) = \frac{1}{9}, \;\; f_1(2/3) = \frac{2}{9}, \;\; f_1(1) = \frac{1}{3}$$

$$f_2(0) = \frac{2}{3}, \quad f_2(1/3) = \frac{7}{9}, \quad f_2(2/3) = \frac{8}{9}, \quad f_2(1) = 1$$

which provides the four (remaining) intervals

$$[0, 1/9], \quad [2/9, 1/3], \quad [2/3, 7/9], \quad [8/9, 1].$$

Next we apply the functions f_1 and f_2 to the points 0, 1/9, 2/9, 1/3, 2/3, 7/9, 8/9, 1 to obtain the next (remaining) eight intervals. Repeating this process up to infinity provides the middle third Cantor set.
(ii) We have

$$f_1(1/2) = 1/6, \quad f_1(1/6) = 1/18, \quad f_1(1/6) = 1/72.$$

The sequence tends to the fixed point $x^* = 0$ of f_1. The sequence

$$f_2(1/2) = 5/6, \quad f_2(5/6) = 17/18, \quad f(17/18) = 53/54$$

tends to the fixed point $x^* = 1$ of f_2.

Problem 3. Consider the iterated function system $F = \{[0,1] : f_1, f_2\}$, where the metric on the unit interval $[0,1]$ is the Euclidean metric and

$$f_1(x) = \frac{1}{2}x, \qquad f_2(x) = \frac{1}{2}x + \frac{1}{2}.$$

The fixed point of f_1 is $x^* = 0$ and the fixed point of f_2 is $x^* = 1$.
(i) Calculate $f_1(1/2)$, $f_1(f_1(1/2))$, $f_1(f_1(f_1(1/2)))$ etc and $f_2(1/2)$, $f_2(f_2(1/2))$, $f_2(f_2(f_2(1/2)))$ etc.
(ii) Find $f_1([0,1]) \cup f_2([0,1])$.

Solution 3. (i) We have

$$f_1(1/2) = 1/4, \quad f_1(f_1(1/2)) = \frac{1}{8}, \quad f_1(f_1(f_1(1/2))) = \frac{1}{16}, \quad \dots$$

The sequence tends to 0 the fixed point of f_1.

$$f_2(1/2) = 3/4, \quad f_2(f_2(1/2)) = \frac{7}{8}, \quad f_2(f_2(f_2(1/2))) = \frac{15}{16}, \quad \dots$$

The sequence tends to 1 the fixed point of f_2.
(ii) We obtain $f_1([0,1]) \cup f_2([0,1]) = [0,1]$.

Problem 4. Show that the construction of the middle third *Cantor set* can be described by the three-dimensional row vector

$$\mathbf{x} = (\, x_1 \quad x_2 \quad x_3 \,) = (\, 1 \quad 0 \quad 1 \,)$$

and repeatedly applying the *Kronecker product* $\otimes$, i.e. calculating

$$\mathbf{x}\otimes\mathbf{x},\ \ \mathbf{x}\otimes\mathbf{x}\otimes\mathbf{x},\ \ \mathbf{x}\otimes\mathbf{x}\otimes\mathbf{x}\otimes\mathbf{x},\ \ \ldots$$

Solution 4. Starting from the unit interval $[0,1]$ the three-dimensional vector $\mathbf{x}$ describes that we partition the unit interval into the three subinterval $[0,1/3]$, $(1/3,2/3)$, $[2/3,1]$. Owing to $x_1=1$ we keep the interval $[0,1/3]$, owing to $x_2=0$ we remove the middle interval $(1/3,2/3)$ and owing to $x_3=1$ we keep the interval $[2/3,1]$. Now $\mathbf{x}\otimes\mathbf{x}$ is the nine dimensional row vector

$$\mathbf{x}\otimes\mathbf{x}=(1\ \ 0\ \ 1\ \ 0\ \ 0\ \ 0\ \ 1\ \ 0\ \ 1)$$

which contains four 1's and five 0's. So we divide the unit interval into the nine subintervals

$$[0,1/9],\ (1/9,2/9),\ [2/9,3/9],$$
$$(3/9,4/9),\ [4/9,5/9],\ (5/9,6/9),$$
$$[6/9,7/9],\ (7/9,8/9),\ [8/9,1]$$

and remove the one where the entry of the nine dimensional vector $\mathbf{x}\otimes\mathbf{x}$ is 0. This means we remove the five intervals

$$(1/9,2/9),\ (3/9,4/9),\ [4/9,5/9],\ (5/9,6/9),\ (7/9,8/9)$$

and we are left with the four intervals

$$[0,1/9],\quad [2/9,3/9],\quad [6/9,7/9],\quad [8/9,1].$$

Now we repeat this procedure for the row vectors

$$\mathbf{x}\otimes\mathbf{x}\otimes\mathbf{x},\quad \mathbf{x}\otimes\mathbf{x}\otimes\mathbf{x}\otimes\mathbf{x},\quad \mathbf{x}\otimes\mathbf{x}\otimes\mathbf{x}\otimes\mathbf{x}\otimes\mathbf{x},\ \ldots$$

Note that for $\mathbf{x}\otimes\mathbf{x}\otimes\mathbf{x}$ we have $2^3=8$ 1's and therefore $3^3-2^3=19$ 0's. This results in the middle third Cantor set.

Problem 5. Consider the unit interval $[0,1]$. The construction of the *Smith-Volterra-Cantor set* starts with the removal of the middle set of length $1/4\equiv 1/2^2$ from $[0,1]$, i.e. we obtain the set

$$[0,3/8]\cup[5/8,1].$$

Next we remove the subintervals of length $1/16\equiv 1/2^4$ from the middle of each of the two remaining intervals $[0,3/8]$ and $[5/8,1]$. We arrive at the four intervals

$$[0,5/32]\cup[7/32,3/8]\cup[5/8,25/32]\cup[27/32,1].$$

Thus we remove subintervals of length $1/2^{2n}$ from the middle of each of the 2^{n-1} remaining intervals. Repeating this process to infinity we obtain the Smith-Volterra-Cantor set. Find the total length of the intervals removed from $[0,1]$.

Solution 5. We have

$$\sum_{k=0}^{\infty} \frac{2^k}{2^{2k+2}} = \frac{1}{4} + \frac{1}{8} + \frac{1}{16} + \cdots + \frac{1}{2^k} + \cdots = \frac{1}{2}$$

with $1 - 1/2 = 1/2$. Thus the set of the remaining points have positive measure of $1/2$.

Problem 6. Let $I = [0,1]$ and $0 < x_0 < x_1 < 1$. A *cookie-cutter map* is a mapping

$$f : [0, x_0] \cup [x_1, 1] \mapsto I$$

with the properties that (i) $f|_{[0,x_0]}$ and $f|_{[x_1,1]}$ are $1-1$ maps onto I, and (ii) f is $C^{1+\gamma}$ differentiable, i.e. differentiable with a Hölder continuous derivative Df satisfying

$$|Df(x) - Df(y)| < c|x-y|^{\gamma}$$

for some $c > 0$ and $|Df(x)| > 1$ for all $x \in [0,x_0] \cup [x_1,1]$. Then the *cookie-cutter set* associated with the map f is the set

$$S := \{ x \in [0,x_0] \cup [x_1,1] \, : \, f^{(n)} \in [0,x_0] \cup [x_1,1] \text{ for all } n \geq 0 \}.$$

Let $x_0 = 1/3$, $x_1 = 2/3$ and $f(x) = 3x \bmod 1$. Find S.

Solution 6. The set S is the middle third Cantor set.

Problem 7. We start with the unit interval $[0,1]$ to construct a two-scale Cantor set. We keep the two intervals $[0,1/4]$ and $[3/5,1]$. So the first interval has length $\ell_1 = 1/4$ and the second interval has length $\ell_2 = 2/5$. The removed interval is $(1/4,3/5)$. The next intervals can be found with the contractive maps $f_1 : [0,1] \to [0,1]$, $f_2 : [0,1] \to [0,1]$

$$f_1(x) = \frac{1}{4}x, \qquad f_2(x) = \frac{2}{5}x + \frac{3}{5}.$$

We have

$$f_1(0) = 0, \;\; f_1(1/4) = \frac{1}{2^4}, \;\; f_1(3/5) = \frac{3}{20}, \;\; f_1(1) = \frac{1}{4}$$

$$f_2(0) = \frac{3}{5}, \;\; f_2(1/4) = \frac{7}{10}, \;\; f_2(3/5) = \frac{21}{25}, \;\; f_2(1) = 1.$$

This provides the remaining intervals $[0,1/4]$, $[3/20,1/4]$, $[3/5,7/10]$, $[21/25,1]$. Repeating this process provides a two-scale Cantor set with $\ell_1 = 1/4$ and $\ell_2 = 2/5$. Find the box counting dimension.

Solution 7. In the n-th generation we have 2^n (non-overlapping) intervals. The shortest interval has length $\ell_1^n = (1/4)^n$ and the longest interval has length $\ell_2^n = (2/5)^n$. There are

$$\binom{n}{k} = \frac{n!}{k!(n-k)!}$$

intervals with length $\ell_1^n \ell_2^{n-k}$ where $k = 0, 1, \ldots, n$. Thus we find the measure μ_d in the n-th generation as

$$\mu_d = \sum_{j=1}^{2^n} \ell_j^d = \sum_{k=0}^{n} \binom{n}{k} \ell_1^{kd} \ell_2^{(n-k)d} = (\ell_1^d + \ell_2^d)^n.$$

With $n \to \infty$ and then $\ell_2^n \to 0$ the quantity μ_d stays finite if and only if $d = D$, where D satisfies the equation

$$\ell_1^D + \ell_2^D = 1$$

with $\ell_1 = 1/4$ and $\ell_2 = 2/5$. The solution of this equation is $D = 0.6110...$.

Problem 8. Consider the *tent map* $f : \mathbb{R} \to \mathbb{R}$ defined by

$$f(x) := \begin{cases} 3x & \text{if } x \le \frac{1}{2} \\ 3 - 3x & \text{if } x > \frac{1}{2} \end{cases}. \tag{1}$$

Note that $f(1/2) = 3/2 > 1$. Discuss the dynamics of the map f on the set Σ of all points x whose iterates stay in the unit interval $[0,1]$. For any other point the iterates approach $-\infty$.

Solution 8. First we note that $f(x) \le 1$ for $x \in [1/3, 2/3]$ and $f(0) = 0$, $f(1) = 0$. The fixed points are $x^* = 0$ and $x^* = 3/4$. If $x \in (1/3, 2/3)$ then the orbit escapes to $-\infty$. We show that Σ is the ternary Cantor set. To prove that Σ is the ternary Cantor set we use the representation of the Cantor set in ternary numbers. We adopt the convention that, where two alternative representations of a given number exist, we choose the form that ends with an infinite string of 2's, rather than the form that ends with a 1 followed by an infinite string of 0's. The Cantor set then consists of the numbers in $[0,1]$ whose ternary expansions contain only 0's and 2's. Let

$$x = 0.x_1 x_2 x_3 \ldots \tag{2}$$

in ternary form and introduce the notation

$$\overline{x}_i := 2 - x_i.$$

Multiplication by 3 shifts x one digit to the left, and $3 = 2.222\ldots$ in ternary form. Hence

$$f(x) = \begin{cases} x_1.x_2x_3x_4\ldots \text{ if } 0 \leq x \leq \frac{1}{2} \\ \overline{x}_1.\overline{x}_2\overline{x}_3\overline{x}_4\ldots \text{ if } \frac{1}{2} < x \leq 1. \end{cases}$$

Note that $\overline{1} = 1$. Hence $f(x) \notin [0,1]$ if $x_1 = 1$. Otherwise, since $\overline{0} = 2$ and $\overline{2} = 0$,

$$f(x) = \begin{cases} 0.x_2x_3x_4\ldots \text{ if } x_1 = 0 \\ 0.\overline{x}_2\overline{x}_3\overline{x}_4\ldots \text{ if } x_1 = 2. \end{cases}$$

Hence $f(x) \in [0,1]$ if and only if $x_1 = 0$ or 2. An inductive argument shows that for all $i \in \mathbb{N}$,

$$f^{(i)}(x) \in [0,1] \quad \text{if and only if} \quad x_i = 0 \text{ or } x_i = 2.$$

Thus $x \in \Sigma$ if and only if x is in the ternary Cantor set and so Σ is the ternary Cantor set. Hence inspection of the formula (1) shows that f maps Σ into itself.

Problem 9. What is the box counting dimension of the countable set S given by the infinite sequence

$$S = \{0,\ 1/2,\ 1/3,\ 1/4, \ldots\}?$$

Solution 9. Covering a set S with a grid of box-size ϵ, the box counting dimension is defined as

$$\dim_B(S) := \lim_{\epsilon \to 0} \frac{\ln(N(\epsilon))}{\ln(1/\epsilon)}$$

where $N(\epsilon)$ is the number of occupied boxes. Consider boxes of some size $0 < \epsilon < 1/2$ and let $k > 1$ be the integer such that

$$\frac{1}{(k-1)k} > \epsilon > \frac{1}{(k+1)k}.$$

For example, $\epsilon = 1/8$ provides $k = 3$. In the finite subset of S

$$\{0,\ 1/2,\ 1/3, \ldots, 1/k\}$$

the two points that are closest are separated by a distance $1/((k-1)/k)$ which is larger than ϵ. Thus one needs at least k boxes to cover the set S. Consequently

$$\frac{\ln(N(\epsilon))}{\ln(1/\epsilon)} \geq \frac{\ln(k)}{\ln(k(k+1))} \to \frac{1}{2}$$

if $k \to \infty$. The box counting dimension $\dim_B(S)$ is thus greater or equal to $1/2$. Analogously with the same k, $(k+1)$ boxes are enough to cover the interval $[0, 1/k]$, and another $(k-1)$ boxes can cover the remaining points. This yields $N(\epsilon) \leq 2k$ and $\dim_B(S) \leq 1/2$. Hence $\dim_B(S) = 1/2$.

Problem 10. A ratio list is a finite list of positive numbers, $(r_1, r_2, \cdots, r_n)$. A ratio list $(r_1, r_2, \cdots, r_n)$ is called contracting (or hyperbolic) iff $r_i < 1$ for all i. The number s is called the *similarity dimension* of a (nonempty compact) set S iff there is a finite "decomposition" of S

$$S = \bigcup_{i=1}^{n} f_i[S]$$

where $(f_1, f_2, \cdots, f_n)$ is an iterated function system of similarities realizing a ratio list with dimension s. The dimension associated with a contracting ratio list $(r_1, r_2, \cdots, r_n)$ is the positive number s such that

$$r_1^s + r_2^s + \cdots + r_n^s = 1.$$

The number s is 0 if and only if $n = 1$. Show that there is only one solution to s.

The Sierpinski gasket is an invariant set for an iterated function system realizing the ratio list $(1/2, 1/2, 1/2)$ with $n = 3$.

Solution 10. Consider the function $\Phi : [0, \infty) \to [0, \infty)$ defined by

$$\Phi(s) = \sum_{i=1}^{n} r_i^s.$$

Then Φ is a continuous function, $\Phi(0) = n \geq 1$ and $\lim_{s \to \infty} \Phi(s) = 0 < 1$. Therefore, by the intermediate value theorem, there is at least one value s with $\Phi(s) = 1$. The derivative of Φ is

$$\sum_{i=1}^{n} r_i^s \log(r_i).$$

This is smaller than 0. Thus Φ is strictly decreasing. Therefore there is only one solution s to $\Phi(s) = 1$. If $n > 1$, then $\Phi(0) > 1$, and therefore $s \neq 0$.

Problem 11. Consider the logistic map $f : \mathbb{R} \to \mathbb{R}$, where

$$f(x) = rx(1-x), \quad r > 4 \tag{1}$$

or

$$x_{t+1} = rx_t(1 - x_t), \qquad r > 4, \quad t = 0, 1, 2, \ldots \tag{2}$$

Assume that $x_0 \in [0, 1]$. Discuss the solutions.

Solution 11. The two fixed points are

$$x_0^* = 0, \qquad x_1^* = \frac{r-1}{r}. \tag{3}$$

Since $r > 4$ we have

$$f\left(\frac{1}{2}\right) > 1$$

Therefore the point $x = 1/2$ is forward asymptotic to infinity. Since $f(1/2) > 1$, $f(1) = 0$, $f(0) = 0$, the intermediate value theorem implies there exist x_0 in $[0, 1/2]$ and x_1 in $[1/2, 1]$ such that $f(x_0) = f(x_1) = 1$. Consequently, $f^{(2)}(x_0) = 0$ and $f^{(2)}(x_1) = 0$ and both x_0 and x_1 are eventually fixed. There are infinitely many points which are eventually fixed at 0 and there are infinitely many points which are eventually fixed at p_r. Next we construct all points of $[0, 1]$ which are forward asymptotic to something in $[0, 1]$. For each $n \in \mathbb{N}$ we define

$$\Lambda_n := \{ x : f^{(n)}(x) \text{ is in } [0, 1] \}.$$

Moreover we define

$$\Lambda := \bigcap_{n=1}^{\infty} \Lambda_n.$$

This is the set of points that remain in $[0, 1]$ forever under iteration of f. The set of real numbers x in $[0, 1]$ satisfying the condition that f is not in $[0, 1]$ is the interval

$$\left(\frac{1}{2} - \frac{\sqrt{r^2 - 4r}}{2r}, \frac{1}{2} + \frac{\sqrt{r^2 - 4r}}{2r} \right)$$

and

$$\Lambda_1 = \left[0, \frac{1}{2} - \frac{\sqrt{r^2 - 4r}}{2r} \right] \bigcup \left[\frac{1}{2} + \frac{\sqrt{r^2 - 4r}}{2r}, 1 \right].$$

The set Λ_n consists of 2^n disjoint closed intervals for all $n \in \mathbb{N}$. If I is one of the closed intervals in Λ_n, then $f^{(n)} : I \to [0, 1]$ is one-to-one and onto.

Consider $f_r(x) = rx(1 - x)$ with $r > 2 + \sqrt{5}$. Then the set Λ is a Cantor set.

Problem 12. Consider the set C obtained from the unit interval $[0, 1]$ by first removing the middle third of the interval and then removing the middle fifths of the two remaining intervals. Now iterate this process, first removing the middle thirds, then removing middle fifths. The set C is what remains when this process is repeated infinitely. Is C a fractal? If so, what is its box counting dimension?

Solution 12. Yes we have a fractal and the box counting dimension of C is $D = \ln(4)/\ln(15/2)$.

Problem 13. Can a fractal that is totally disconnected (topological dimension 0) have a fractal dimension larger than 1?

Solution 13. Yes. Consider the Cantor set that we obtain as the Cartesian product of two middle fifths set. Then the fractal dimension would be

$$D = \ln(4)/\ln(5/2) = 2\ln(2)/\ln(5/2) = 2 \cdot 0.75647... > 1.$$

Problem 14. Give the matrices to perform any one of the following operations

- Rotation around the x_3-axis and the other two axis
- Scaling
- Shearing of the x_1 by the x_3-coordinate
- Translation

in any order using only matrix multiplications. Use the homogeneous form of a point $(x_1, x_2, x_3, 1)^T$ and 4×4 matrices to solve the problem.

Solution 14. The *rotation matrix* for rotation around the x_3-axis by an angle θ is given by

$$R_3(\theta) = \begin{pmatrix} \cos(\theta) & -\sin(\theta) & 0 & 0 \\ \sin(\theta) & \cos(\theta) & 0 & 0 \\ 0 & 0 & 1 & 0 \\ 0 & 0 & 0 & 1 \end{pmatrix}.$$

For the other two axis we have

$$R_1(\theta) = \begin{pmatrix} 1 & 0 & 0 & 0 \\ 0 & \cos(\theta) & -\sin(\theta) & 0 \\ 0 & \sin(\theta) & \cos(\theta) & 0 \\ 0 & 0 & 0 & 1 \end{pmatrix}$$

$$R_2(\theta) = \begin{pmatrix} \cos(\theta) & 0 & -\sin(\theta) & 0 \\ 0 & 1 & 0 & 0 \\ \sin(\theta) & 0 & \cos(\theta) & 0 \\ 0 & 0 & 0 & 1 \end{pmatrix}$$

Scaling is achieved by

$$S = \begin{pmatrix} S_1 & 0 & 0 & 0 \\ 0 & S_2 & 0 & 0 \\ 0 & 0 & S_3 & 0 \\ 0 & 0 & 0 & 1 \end{pmatrix}$$

Shearing the x_1-coordinate by the x_3-coordinate is given by

$$H_{13} = \begin{pmatrix} 1 & 0 & s & 0 \\ 0 & 1 & 0 & 0 \\ 0 & 0 & 1 & 0 \\ 0 & 0 & 0 & 1 \end{pmatrix}.$$

Finally *translation* is represented by

$$T = \begin{pmatrix} 1 & 0 & 0 & T_1 \\ 0 & 1 & 0 & T_2 \\ 0 & 0 & 1 & T_3 \\ 0 & 0 & 0 & 1 \end{pmatrix}.$$

Now any sequence of operations can be combined by multiplying the corresponding matrices together. This matrix can then be applied to a point to transform the point.

Problem 15. Compute exactly the area of the *Koch snowflake*. Koch's snowflake may be enclosed within a rectangle having length 1 and width $2\sqrt{3}/3$.

Solution 15. Hence its area is not larger than $2\sqrt{3}/3$. Let $A(0)$ be the area of the equilateral triangle with sides of length 1. Since the height of this triangle is $\sqrt{3}/2$, we have that $A(0) = \sqrt{3}/4$. Now at the k-th step in the construction of the Koch snowflake, $3 \cdot 4^{k-1}$ equilateral triangles each with area $\sqrt{3}/(4 \cdot 3^{2k})$ are added to the snowflake. The combined area of these triangles is

$$3 \cdot 4^{k-1} \cdot \sqrt{3}/(4 \cdot 3^{2k}) = \frac{3}{4} \cdot 4^k \cdot \sqrt{3}/(4 \cdot 9^k) = \frac{3}{16}\sqrt{3}(4/9)^k.$$

Thus the area B of all these triangles is the sum from $k = 0$ to ∞ of the previous expression. Therefore

$$B = \frac{3}{16}\frac{4}{5}\sqrt{3} = \frac{3}{20}\sqrt{3}.$$

Hence the area A of the Koch snowflake is

$$A = \frac{1}{4}\sqrt{3} + \frac{3}{20}\sqrt{3} = \frac{2}{5}\sqrt{3}.$$

Problem 16. Describe the filled *Julia set* for the map $f : \mathbb{C} \to \mathbb{C}$, $f(z) = z^3$. The fixed points of f are 0, +1, −1.

Solution 16. The filled Julia set for $f(z) = z^3$ is the closed unit disk. We have

$$f(z) = z^3, \quad f^{(2)} = z^9, \quad \dots \quad f^{(n)} = z^{3^n}.$$

Now we have

$$z^k = r^k(\cos(k\theta) + i\sin(k\theta)) \equiv r^k e^{ik\theta}$$

where $r = |z|$ and $\tan(\theta) = \Im(z)/\Re(z)$. Thus we have

$$f^{(n)}(z) = z^{3^n} = r^{3^n} e^{i3^n\theta}.$$

It follows that $|z| = 1$ implies that $|f(z)| = 1$, that is the unit circle is invariant, since

$$|f(z)| = |z^3| = |z|^3.$$

Thus if $|z| = 1$ we have $|f(z)| = 1$.

Problem 17. Consider the map

$$g_\lambda(z) = r(z - z^3), \quad r > 0.$$

(i) Find the fixed points.
(ii) Show that the points

$$p_+(\lambda) = \sqrt{\frac{r+1}{r}}, \quad p_-(\lambda) = -\sqrt{\frac{r+1}{r}}$$

lie on a cycle of period 2.

Solution 17. (i) We find the three fixed points

$$z_1^* = 0, \quad z_{2,3}^* = \pm\sqrt{\frac{r-1}{r}}.$$

(ii) We have

$$g_\lambda(z = p_+) = p_-, \qquad g_\lambda(z = p_-) = p_+.$$

Problem 18. The observed volume $V(\varepsilon)$ of a fractal in dimensions d when covered with N d-dimensional cubes has the form

$$V(\varepsilon) = \varepsilon^d N(\varepsilon).$$

Since $N(\varepsilon) \sim \varepsilon^{D_0}$, this implies

$$V(\varepsilon) = \varepsilon^{d-D_0}$$

where $d - D_0$ is called the co-dimension of the fractal. If $d - D_0 > 0$, the observed volume decreases with resolution ε. This defines the category of *thin* fractals, for which $V(\varepsilon)$ vanishes for $\varepsilon \to 0$. If $d - D_0 = 0$, this does not imply that we have a traditional geometric object. We can have a ramified structure. Such objects converge to a nonzero volume. Such objects are called *fat fractals*. Fat fractals can be distinguished by measuring the difference between observed and the real volume by means of an exponent α as

$$V(\varepsilon) - V \sim \varepsilon^{\alpha}.$$

For fat fractals we have $0 < \alpha < 1$.
Consider the unit interval $[0, 1]$. Remove instead of the middle third each time from the unit interval $[0, 1]$ the proportions

$$1/3, \quad 1/9 = 1/3^2, \quad 1/27 = 3^3, \quad \ldots.$$

What is the dimension of the fat fractal?

Solution 18. The volume in the n'th construction step is

$$V_n = \prod_{j=1}^{n} (1 - 3^{-j})$$

which converges towards a nonzero value of $V \approx 0.560$. The logarithmic difference $V_n - V$ can be expanded as follows

$$\ln(V_n - V) = \ln(V_n) + \ln(1 - \prod_{j=n+1}^{\infty} (1 - 3^{-j})).$$

For n sufficiently large, the second term can be approximated by

$$\prod_{j=n+1}^{\infty} (1 - 3^{-j}) \approx 1 - \sum_{j=n+1}^{\infty} 3^{-j} = 1 - 3^{-n}/2.$$

The smallest distance occurring in the n'th step is

$$\varepsilon(n) = 3^{-n} 2^{-n+1} \prod_{j=1}^{n-1} (1 - 3^{-j})$$

located between intervals of length $2^{-n}\prod_{j=1}^{n}(1-3^{-j})$. Hence $\varepsilon(n)$ should be taken for a covering of the structure. For large n, it can be seen that $\ln(\varepsilon(n)) \approx -n\ln(6)$. It follows that

$$\alpha = \frac{\ln(V_n - V)}{\ln(\varepsilon(n))} \approx \frac{\ln(V_n) - n\ln(3) - \ln(2)}{n\ln(1/6)}$$

which yields $\alpha = \frac{\ln(3)}{\ln(6)}$.

Problem 19. Each $x \in [0,1]$ can be written as

$$x = \sum_{j=1}^{\infty} \frac{\epsilon_j}{2^j}$$

with $\epsilon_j = 0$ or $\epsilon_j = 1$. Define the function $f : [0,1] \to [0,1)$ as

$$f(x) = \sum_{j=1}^{\infty} \frac{2\epsilon_j}{3^j}.$$

The function f is known as the *Cantor function.* Let $x = 1/8$. Find the ϵ_j's and then $f(x)$.

Solution 19. For $x = 1/8 = 1/2^3$ we find that $\epsilon_3 = 1$ and all other ϵ_j's are 0. Thus

$$f(1/8) = \frac{2}{3^3} = \frac{2}{27}.$$

Problem 20. The *Koch curve* is self-similar on each length scale. Each time the length of the unit is reduced by a factor of 3, the number of segments is increased by a factor of 4. Find the box counting dimension D.

Solution 20. We have $N = 4$ and $\ell = 1/3$. From $N = 1/\ell^D$ we arrive at

$$D = \frac{\ln(N)}{\ln(1/\ell)} = \frac{\ln(4)}{\ln(3)} \approx 1.2619...$$

Problem 21. Let $n \geq 2$. Given an $n \times n$ binary matrix $A = (a_{jk})$ with $j,k = 0,1,\ldots,n-1$. Consider the following partition of the unit square $[0,1]\times[0,1]$. With every entry a_{jk} ($a_{jk} \in \{0,1\}$) in the matrix A we associate a square of size $1/n \times 1/n$ at position (vertex) $(j/n, k/n)$ in the unit square with the square given by $([j/n,(j+1)/n] \times [k/n,(k+1)/n]$. If

$a_{jk} = 1$ we colour the a_{jk}'s corresponding black otherwise white. Apply it to the symmetric binary 3×3 matrix

$$A = \begin{pmatrix} 1 & 0 & 1 \\ 0 & 1 & 0 \\ 1 & 0 & 1 \end{pmatrix}$$

and $A \otimes A$ given by the 9×9 symmetric matrix

$$A \otimes A = \begin{pmatrix} 1 & 0 & 1 & 0 & 0 & 0 & 1 & 0 & 1 \\ 0 & 1 & 0 & 0 & 0 & 0 & 0 & 1 & 0 \\ 1 & 0 & 1 & 0 & 0 & 0 & 1 & 0 & 1 \\ 0 & 0 & 0 & 1 & 0 & 1 & 0 & 0 & 0 \\ 0 & 0 & 0 & 0 & 1 & 0 & 0 & 0 & 0 \\ 0 & 0 & 0 & 1 & 0 & 1 & 0 & 0 & 0 \\ 01 & 0 & 1 & 0 & 0 & 0 & 1 & 0 & 1 \\ 0 & 1 & 0 & 0 & 0 & 0 & 0 & 1 & 0 \\ 1 & 0 & 1 & 0 & 0 & 0 & 1 & 0 & 1 \end{pmatrix}$$

where $\otimes$ denotes the Kronecker product.

Solution 21. For A and $A \otimes A$ we obtain the figure

Problem 22. Starting from the unit square the figure shows the successive stages of generating a fractal. Find the box counting dimension (capacity).

Solution 22. At step $k \geq 1$ we have 5^k rectangles of length $(1/3)^k$ each. Thus

$$\frac{\ln(5^k)}{\ln(3^k)} = \frac{k \ln(5)}{k \ln(3)} = \frac{\ln(5)}{\ln(3)} = 1.46497...$$

Problem 23. The *Sierpinski carpet* is constructed as follows: Consider the unit square $[0, 1] \times [0, 1]$. Partition the unit square into nine equal squares by removing the interior of the middle one. This process is then repeated in each of the remaining eight squares. The first three steps in this construction are displayed in the figure. Show that the fractal dimension of the Sierpinski carpet is $\log_3(8)$. Find the area of the *Sierpinski carpet* starting from the unit square $[0,1]^2$.

Solution 23. The area of the unit square is $A(0) = 1$. To obtain $A(k+1)$ from $A(k)$ we scale $A(k)$ by $1/3$. This reduces the area by $(1/3)^2 = 1/9$. However we arrive at 8 copies of the scaled version to form $A(k+1)$. Hence the area of $A(k+1)$ is given by $(8/9)$th of the area of $A(k)$ and consequently the area of $A(n)$ is $(8/9)^n$. Therefore

$$\lim_{n\to\infty} A(n) = 0.$$

Problem 24. Starting from the unit square the figure shows the successive stages of generating a fractal, i.e. we remove iteratively a square of relative area $\frac{1}{9}$ from the center of a square. Find the box counting dimension (capacity).

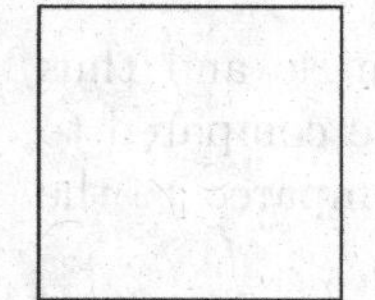 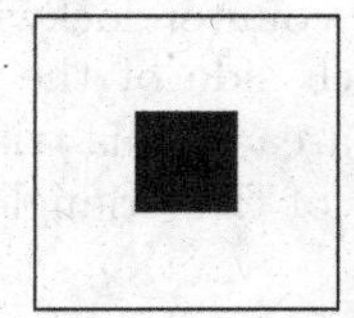 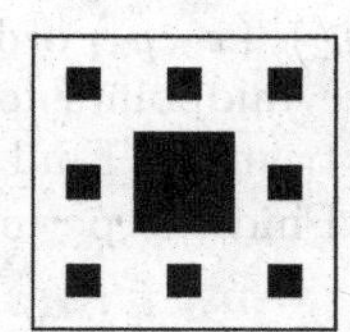

Solution 24. At step $k = 0$ we have one white square of side length $\epsilon = 1$. For step $k = 1$ we have $N = 8$ white squares of side length $1/3$. At step $k = 2$ we have $N = 8 \cdot 8 = 64$ white squares each of side length $1/3^2 = 1/9$. At the nth step we have $N = 8^n$ white squares of side length $\epsilon = 1/3^n$. Thus the box counting dimension is

$$d = \lim_{n\to\infty} \frac{\ln(8^n)}{\ln(3^n)} = \lim_{n\to\infty} \frac{n\ln(2^3)}{n\ln(3)} = \frac{3\ln(2)}{\ln(3)} \approx 1.893$$

Problem 25. Consider the 3×3 binary matrix

$$A = \begin{pmatrix} 1 & 1 & 1 \\ 1 & 0 & 1 \\ 1 & 1 & 1 \end{pmatrix}.$$

We identify each entry of the matrix with a square of size $1/3 \times 1/3$. and identify 0 with a black square and 1 with a white square. Thus we have 8 white squares and 1 black square. Let $\otimes$ be the Kronecker product. What fractal is generated by $A \otimes A$, $A \otimes A \otimes A$, etc.

Solution 25. The $3^2 \times 3^2$ binary matrix $A \otimes A$ has $9^2 = 81$ squares of size $1/9 \times 1/9$ with $8^2 = 64$ white squares and $4^2 + 1 = 17$ black squares.

$$A \otimes A = \begin{pmatrix} 1 & 1 & 1 & 1 & 1 & 1 & 1 & 1 & 1 \\ 1 & 0 & 1 & 0 & 1 & 1 & 1 & 0 & 1 \\ 1 & 1 & 1 & 1 & 1 & 1 & 1 & 1 & 1 \\ 1 & 1 & 1 & 0 & 0 & 0 & 1 & 1 & 1 \\ 1 & 1 & 1 & 0 & 0 & 0 & 1 & 1 & 1 \\ 1 & 1 & 1 & 0 & 0 & 0 & 1 & 1 & 1 \\ 1 & 1 & 1 & 1 & 1 & 1 & 1 & 1 & 1 \\ 1 & 0 & 1 & 1 & 0 & 1 & 1 & 0 & 1 \\ 1 & 1 & 1 & 1 & 1 & 1 & 1 & 1 & 1 \end{pmatrix}.$$

Then $A \otimes A \otimes \cdots \otimes A$ (n-factors) we have a $3^n \times 3^n$ matrix with 8^n white squares. Thus we generate the fractal of the previous problem.

Problem 26. Consider a triangle in the Euclidean plane $\mathbb{R}^2$ with the vertices (x_1, y_1), (x_2, y_2), (x_3, y_3) ordered counterclockwise.
(i) Consider the three midpoints for each side of the triangle and thus construct a midpoint triangle. Find the area of this triangle compared to the original triangle. Find the perimeter of this triangle compared to the original triangle.
(ii) The area A_1 of a triangle with vertices (x_1, y_1), (x_2, y_2), (x_3, y_3) is given by

$$A_1 = \frac{1}{2} \det \begin{pmatrix} x_1 & y_1 & 1 \\ x_2 & y_2 & 1 \\ x_3 & y_3 & 1 \end{pmatrix} = \frac{1}{2}(x_2 y_3 - x_3 y_2 + x_1 y_2 - x_2 y_1 + x_3 y_1 - x_1 y_3).$$

Let $s_1, s_2 \in (0,1)$ and $s_1 + s_2 = 1$. Find the area of the triangle with the vertices

$$(s_1 x_1 + s_2 x_2, s_1 y_1 + s_2 y_2), \quad (s_1 x_2 + s_2 x_3, s_1 y_2 + s_2 y_3), \quad (s_1 x_3 + s_2 x_1, s_1 y_3 + s_2 y_1).$$

Solution 26. (i) The area is $\frac{1}{4}$ of the original triangle. The perimeter is $\frac{1}{2}$ of the original triangle.
(ii) We have

$$A_2 = \frac{1}{2} \det \begin{pmatrix} s_1 x_1 + s_2 x_2 & s_1 y_1 + s_2 y_2 & 1 \\ s_1 x_2 + s_2 x_3 & s_1 y_2 + s_2 y_3 & 1 \\ s_1 x_3 + s_2 x_1 & s_1 y_3 + s_2 y_1 & 1 \end{pmatrix} = (s_1^2 + s_2^2 - s_1 s_2) A_1.$$

Hence for $s_1 = s_2 = 1/2$ we have $A_2 = \frac{1}{4}A_1$.

Problem 27. Find the box counting dimension of the fractal generated by the process displayed in the figure

  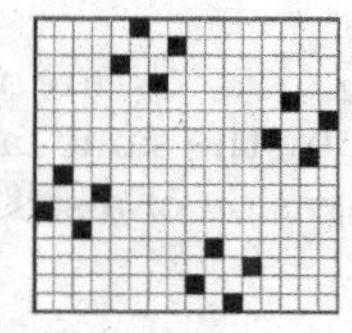

Solution 27. At the nth step we cover the set with a grid of black boxes of size $1/4^n$. The number of boxes is $N(\epsilon) = 4^n$. Thus

$$d = \lim_{\epsilon \to 0} \frac{\ln(N(\epsilon))}{\ln(1/\epsilon)} = \lim_{k \to \infty} \frac{\ln(4^n)}{\ln(4^n)} = 1.$$

Problem 28. Consider the 4×4 permutation matrix

$$P = \begin{pmatrix} 0 & 1 & 0 & 0 \\ 0 & 0 & 0 & 1 \\ 1 & 0 & 0 & 0 \\ 0 & 0 & 1 & 0 \end{pmatrix}.$$

What fractal is generated by $P \otimes P$, $P \otimes P \otimes P$, ... ?

Solution 28. We obtain the fractal from the previous problem.

Problem 29. Find the box counting dimension of the fractal generated by the process displayed in the figure.

Solution 29. For step $k = 0$ we have $N = 1$ and $\epsilon = 1$. For step $k = 1$ we have $\epsilon = 1/3$ and $N = 4$. For step $k = 2$ we have $\epsilon = 1/3^2$ and $N = 4^2$. For the step $k = n$ we have $\epsilon = 1/3^n$ with $N = 4^n$. Thus

$$d = \lim_{n \to \infty} \frac{\ln(4^n)}{\ln(3^n)} = \lim_{n \to \infty} \frac{n \ln(2^2)}{n \ln(3)} = \frac{2 \ln(2)}{\ln(3)} \approx 1.2619$$

Problem 30. Let $k \geq 2$ and $\cup$ be the union of sets. Consider a closed and bounded subset S of the Euclidean space $\mathbb{R}^2$. It is said to be self-similar if it can be expressed in the form

$$S = S_1 \cup S_2 \cup \cdots \cup S_k$$

where the sets S_1, S_2, ..., S_k are nonoverlapping sets, each of which is congruent to S scaled by the same factor s with $0 < s < 1$ (scaling factor). The box counting dimension of a self-similar set S is denoted by $d_B(S)$ and defined by

$$d_B(S) = \frac{\ln(k)}{\ln(1/s)}.$$

(i) Consider the unit square $[0,1]^2$, $k = 4$ and $s = 1/2$. Find $d_B(S)$.
(ii) For the Sierpinski carpet one has $k = 8$ and $s = 1/3$. Find $d_B(S)$.

Solution 30. (i) We have $d_B(S) = 2$
(ii) We have $d_B(S) = 3\ln(2)/\ln(3)$.

Problem 31. The *rotation matrix*

$$R(\theta) := \begin{pmatrix} \cos(\theta) & -\sin(\theta) \\ \sin(\theta) & \cos(\theta) \end{pmatrix}$$

is an element of the Lie group $SO(2,\mathbb{R})$ with the inverse matrix given by

$$R^{-1}(\theta) = R(-\theta) = \begin{pmatrix} \cos(\theta) & \sin(\theta) \\ -\sin(\theta) & \cos(\theta) \end{pmatrix}.$$

Consider the Euclidean space $\mathbb{R}^2$. A *similitude* with a scaling factor $0 < s < 1$ is a map of $\mathbb{R}^2$ into $\mathbb{R}^2$ given by

$$\mathbf{f}(x_1,x_2) = \begin{pmatrix} f_1(x_1,x_2) \\ f_2(x_1,x_2) \end{pmatrix} = sR(\theta)\begin{pmatrix} x_1 \\ x_2 \end{pmatrix} + \begin{pmatrix} t_1 \\ t_2 \end{pmatrix}$$

where $\theta, t_1, t_2 \in \mathbb{R}$. Thus the map consists of a scaling by a factor s, a rotation about the origin $(0,0)$ and a translation in the x_1 direction by t_1 and in the x_2 direction by t_2. Let $\theta = \pi/4$, $s = 1/2$, $t_1 = 1$, $t_2 = -1$. Find

$$\mathbf{f}(0,0), \quad \mathbf{f}(0,1), \quad \mathbf{f}(1,0), \quad \mathbf{f}(1,1).$$

Solution 31. We have the map

$$\mathbf{f}(x_1,x_2) = \frac{1}{2}\begin{pmatrix} 1/\sqrt{2} & -1/\sqrt{2} \\ 1/\sqrt{2} & 1/\sqrt{2} \end{pmatrix}\begin{pmatrix} x_1 \\ x_2 \end{pmatrix} + \begin{pmatrix} t_1 \\ t_2 \end{pmatrix}.$$

Thus

$$\mathbf{f}(0,0) = \begin{pmatrix} 1 \\ -1 \end{pmatrix}, \qquad \mathbf{f}(0,1) = \begin{pmatrix} 1 - 1/(2\sqrt{2}) \\ -1 + 1/(2\sqrt{2}) \end{pmatrix}$$

$$\mathbf{f}(1,0) = \begin{pmatrix} 1 + 1/(2\sqrt{2}) \\ -1 + 1/(2\sqrt{2}) \end{pmatrix}, \qquad \mathbf{f}(1,1) = \begin{pmatrix} 1 + 1/\sqrt{2} \\ -1 + 1/\sqrt{2} \end{pmatrix}.$$

Problem 32. Consider the unit square $[0,1]^2 = [0,1] \times [0,1]$. The Sierpinski carpet is constructed from the eight maps

$$\mathbf{f}_1(x_1,x_2) = \frac{1}{3}\begin{pmatrix} 1 & 0 \\ 0 & 1 \end{pmatrix}\begin{pmatrix} x_1 \\ x_2 \end{pmatrix}$$

$$\mathbf{f}_2(x_1,x_2) = \frac{1}{3}\begin{pmatrix} 1 & 0 \\ 0 & 1 \end{pmatrix}\begin{pmatrix} x_1 \\ x_2 \end{pmatrix} + \begin{pmatrix} 0 \\ 1/3 \end{pmatrix}$$

$$\mathbf{f}_3(x_1,x_2) = \frac{1}{3}\begin{pmatrix} 1 & 0 \\ 0 & 1 \end{pmatrix}\begin{pmatrix} x_1 \\ x_2 \end{pmatrix} + \begin{pmatrix} 0 \\ 2/3 \end{pmatrix}$$

$$\mathbf{f}_4(x_1,x_2) = \frac{1}{3}\begin{pmatrix} 1 & 0 \\ 0 & 1 \end{pmatrix}\begin{pmatrix} x_1 \\ x_2 \end{pmatrix} + \begin{pmatrix} 1/3 \\ 0 \end{pmatrix}$$

$$\mathbf{f}_5(x_1,x_2) = \frac{1}{3}\begin{pmatrix} 1 & 0 \\ 0 & 1 \end{pmatrix}\begin{pmatrix} x_1 \\ x_2 \end{pmatrix} + \begin{pmatrix} 1/3 \\ 2/3 \end{pmatrix}$$

$$\mathbf{f}_6(x_1,x_2) = \frac{1}{3}\begin{pmatrix} 1 & 0 \\ 0 & 1 \end{pmatrix}\begin{pmatrix} x_1 \\ x_2 \end{pmatrix} + \begin{pmatrix} 2/3 \\ 0 \end{pmatrix}$$

$$\mathbf{f}_7(x_1,x_2) = \frac{1}{3}\begin{pmatrix} 1 & 0 \\ 0 & 1 \end{pmatrix}\begin{pmatrix} x_1 \\ x_2 \end{pmatrix} + \begin{pmatrix} 2/3 \\ 1/3 \end{pmatrix}$$

$$\mathbf{f}_8(x_1,x_2) = \frac{1}{3}\begin{pmatrix} 1 & 0 \\ 0 & 1 \end{pmatrix}\begin{pmatrix} x_1 \\ x_2 \end{pmatrix} + \begin{pmatrix} 2/3 \\ 2/3 \end{pmatrix}.$$

Apply the maps to the four vertices of the unit square

$$\mathbf{v}_1 = (x_1,x_2) = (0,0), \quad \mathbf{v}_2 = (x_1,x_2) = (0,1),$$

$$\mathbf{v}_3 = (x_1,x_2) = (1,0), \quad \mathbf{v}_4 = (x_1,x_2) = (1,1).$$

Discuss.

Solution 32. For $\mathbf{f}_1$ we find

$$\mathbf{f}_1(\mathbf{v}_1) = \begin{pmatrix} 0 \\ 0 \end{pmatrix}, \quad \mathbf{f}_1(\mathbf{v}_2) = \begin{pmatrix} 0 \\ 1/3 \end{pmatrix}, \quad \mathbf{f}_1(\mathbf{v}_3) = \begin{pmatrix} 1/3 \\ 0 \end{pmatrix}, \quad \mathbf{f}_1(\mathbf{v}_4) = \begin{pmatrix} 1/3 \\ 1/3 \end{pmatrix}.$$

Consequently we generate the vertices of 8 new squares each with length 1/3.

Problem 33. Consider *Pascal's triangle*

```
                1
               1 1
              1 2 1
             1 3 3 1
            1 4 6 4 1
          1 5 10 10 5 1
        1 6 15 20 15 6 1
      1 7 21 35 35 21 7 1
    1 8 28 56 70 56 28 8 1
```

Apply the mod 2 operation. Discuss.

Solution 33. Applying the mod 2 operation we find

```
        1
       1 1
      1 0 1
     1 1 1 1
    1 0 0 0 1
   1 1 0 0 1 1
  1 0 1 0 1 0 1
 1 1 1 1 1 1 1 1
1 0 0 0 0 0 0 0 1
```

Thus we obtain a Sierpinski triangle.

Problem 34. The picture shows the first three steps in the construction of the *Hironaka curve.*

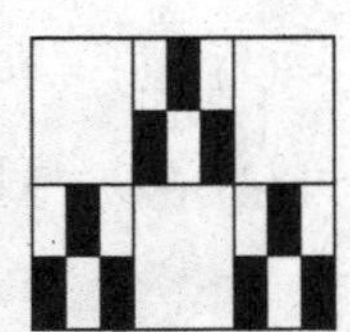
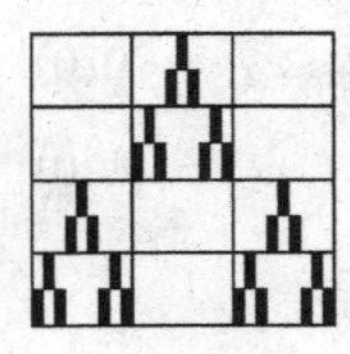

Show that the Kronecker product $\otimes$ and the 2×3 matrix

$$A = \begin{pmatrix} 0 & 1 & 0 \\ 1 & 0 & 1 \end{pmatrix}$$

can be used to construct the Hironaka curve. Here 1 is identified with the black rectangle and 0 with the white rectangle.

Solution 34. For $A \otimes A$ we have the 4×9 matrix

$$A \otimes A = \begin{pmatrix} 0 & 0 & 0 & 0 & 1 & 0 & 0 & 0 & 0 \\ 0 & 0 & 0 & 1 & 0 & 1 & 0 & 0 & 0 \\ 0 & 1 & 0 & 0 & 0 & 0 & 0 & 1 & 0 \\ 1 & 0 & 1 & 0 & 0 & 0 & 1 & 0 & 1 \end{pmatrix}$$

for the second step in the construction of the Hironaka curve. For

$$A \otimes A \otimes \cdots \otimes A, \quad (n - \text{times})$$

we have a $2^n \times 3^n$ matrix.

Problem 35. Consider the iterated function system for the *Hironaka curve*

$$\mathbf{f}_1(\mathbf{x}) = \begin{pmatrix} 1/3 & 0 \\ 0 & 1/2 \end{pmatrix} \begin{pmatrix} x_1 \\ x_2 \end{pmatrix}$$
$$\mathbf{f}_2(\mathbf{x}) = \begin{pmatrix} 1/3 & 0 \\ 0 & 1/2 \end{pmatrix} \begin{pmatrix} x_1 \\ x_2 \end{pmatrix} + \begin{pmatrix} 1/3 \\ 1/2 \end{pmatrix}$$
$$\mathbf{f}_3(\mathbf{x}) = \begin{pmatrix} 1/3 & 0 \\ 0 & 1/2 \end{pmatrix} \begin{pmatrix} x_1 \\ x_2 \end{pmatrix} + \begin{pmatrix} 2/3 \\ 0 \end{pmatrix}$$

and $I_2 = [0,1] \times [0,1]$. Note that the scaling factors in the matrix are not the same. Let

$$\mathbf{x}_1 = \begin{pmatrix} 0 \\ 0 \end{pmatrix}, \quad \mathbf{x}_2 = \begin{pmatrix} 1 \\ 0 \end{pmatrix}, \quad \mathbf{x}_3 = \begin{pmatrix} 0 \\ 1 \end{pmatrix}, \quad \mathbf{x}_4 = \begin{pmatrix} 1 \\ 1 \end{pmatrix}.$$

Apply $\mathbf{f}_1$, $\mathbf{f}_2$, $\mathbf{f}_3$ to the four vectors $\mathbf{x}_1$, $\mathbf{x}_2$, $\mathbf{x}_3$, $\mathbf{x}_4$.

Solution 35. This is the first step in the construction of the Hironaka curve. We have

$$\mathbf{f}_1(\mathbf{x}_1) = \begin{pmatrix} 0 \\ 0 \end{pmatrix}, \ \mathbf{f}_1(\mathbf{x}_2) = \begin{pmatrix} 1/3 \\ 0 \end{pmatrix}, \ \mathbf{f}_1(\mathbf{x}_3) = \begin{pmatrix} 0 \\ 1/2 \end{pmatrix}, \ \mathbf{f}_1(\mathbf{x}_4) = \begin{pmatrix} 1/3 \\ 1/2 \end{pmatrix}$$

$$\mathbf{f}_2(\mathbf{x}_1) = \begin{pmatrix} 1/3 \\ 1/2 \end{pmatrix}, \ \mathbf{f}_2(\mathbf{x}_2) = \begin{pmatrix} 2/3 \\ 1/2 \end{pmatrix}, \ \mathbf{f}_2(\mathbf{x}_3) = \begin{pmatrix} 1/3 \\ 1 \end{pmatrix}, \ \mathbf{f}_2(\mathbf{x}_4) = \begin{pmatrix} 2/3 \\ 1 \end{pmatrix}$$

$$\mathbf{f}_3(\mathbf{x}_1) = \begin{pmatrix} 2/3 \\ 0 \end{pmatrix}, \ \mathbf{f}_3(\mathbf{x}_2) = \begin{pmatrix} 1 \\ 0 \end{pmatrix}, \ \mathbf{f}_3(\mathbf{x}_3) = \begin{pmatrix} 2/3 \\ 1/2 \end{pmatrix}, \ \mathbf{f}_3(\mathbf{x}_4) = \begin{pmatrix} 1 \\ 1/2 \end{pmatrix}.$$

Note that

$$\mathbf{f}_1(\mathbf{x}_4) = \mathbf{f}_2(\mathbf{x}_1) = \begin{pmatrix} 1/3 \\ 1/2 \end{pmatrix}, \qquad \mathbf{f}_2(\mathbf{x}_2) = \mathbf{f}_3(\mathbf{x}_3) = \begin{pmatrix} 2/3 \\ 1/3 \end{pmatrix}.$$

Problem 36. The iterated function for the *von Koch curve* is given by the four functions

$$\mathbf{f}_1(\mathbf{x}) = \frac{1}{3}\begin{pmatrix} 1 & 0 \\ 0 & 1 \end{pmatrix}\begin{pmatrix} x_1 \\ x_2 \end{pmatrix}, \quad \mathbf{f}_2(\mathbf{x}) = \frac{1}{3}\begin{pmatrix} 1/2 & -\sqrt{3}/2 \\ \sqrt{3}/2 & 1/2 \end{pmatrix}\begin{pmatrix} x_1 \\ x_2 \end{pmatrix} + \begin{pmatrix} 1/3 \\ 0 \end{pmatrix}$$

$$\mathbf{f}_3(\mathbf{x}) = \frac{1}{3}\begin{pmatrix} 1/2 & \sqrt{3}/2 \\ -\sqrt{3}/2 & 1/2 \end{pmatrix}\begin{pmatrix} x_1 \\ x_2 \end{pmatrix} + \begin{pmatrix} 1/2 \\ 1/(2\sqrt{3}) \end{pmatrix},$$

$$\mathbf{f}_4(\mathbf{x}) = \frac{1}{3}\begin{pmatrix} 1 & 0 \\ 0 & 1 \end{pmatrix}\begin{pmatrix} x_1 \\ x_2 \end{pmatrix} + \begin{pmatrix} 2/3 \\ 0 \end{pmatrix}$$

with $x_1 \in [0,1]$ and $x_2 \in [0, 1/(2\sqrt{3})]$. Apply the functions to the five vectors

$$\mathbf{x}_1 = \begin{pmatrix} 0 \\ 0 \end{pmatrix}, \quad \mathbf{x}_2 = \begin{pmatrix} 1/3 \\ 0 \end{pmatrix}, \quad \mathbf{x}_3 = \begin{pmatrix} 1/2 \\ 1/(2\sqrt{3}) \end{pmatrix}, \quad \mathbf{x}_4 = \begin{pmatrix} 2/3 \\ 0 \end{pmatrix}, \quad \mathbf{x}_5 = \begin{pmatrix} 1 \\ 0 \end{pmatrix}$$

to find the vectors (positions) of the second step in the construction of the Koch curve.

Solution 36. We have

$$\mathbf{f}_1(\mathbf{x}_1) = \begin{pmatrix} 0 \\ 0 \end{pmatrix}, \quad \mathbf{f}_1(\mathbf{x}_2) = \begin{pmatrix} 1/9 \\ 0 \end{pmatrix}, \quad \mathbf{f}_1(\mathbf{x}_3) = \begin{pmatrix} 1/6 \\ 1/(6\sqrt{3}) \end{pmatrix},$$

$$\mathbf{f}_1(\mathbf{x}_4) = \begin{pmatrix} 2/9 \\ 0 \end{pmatrix}, \quad \mathbf{f}_1(\mathbf{x}_5) = \begin{pmatrix} 1/3 \\ 0 \end{pmatrix}$$

$$\mathbf{f}_2(\mathbf{x}_1) = \begin{pmatrix} 1/3 \\ 0 \end{pmatrix}, \quad \mathbf{f}_2(\mathbf{x}_2) = \begin{pmatrix} 7/18 \\ 1/(6\sqrt{3}) \end{pmatrix}, \quad \mathbf{f}_2(\mathbf{x}_3) = \begin{pmatrix} 1/3 \\ 1/(3\sqrt{3}) \end{pmatrix},$$

$$\mathbf{f}_2(\mathbf{x}_4) = \begin{pmatrix} 7/9 \\ 1/(3\sqrt{3}) \end{pmatrix}, \quad \mathbf{f}_2(\mathbf{x}_5) = \begin{pmatrix} 1/2 \\ 1/(2\sqrt{3}) \end{pmatrix}$$

$$\mathbf{f}_3(\mathbf{x}_1) = \begin{pmatrix} 1/2 \\ 1/(2\sqrt{3}) \end{pmatrix}, \quad \mathbf{f}_3(\mathbf{x}_2) = \begin{pmatrix} 5/9 \\ 1/(3\sqrt{3}) \end{pmatrix}, \quad \mathbf{f}_3(\mathbf{x}_3) = \begin{pmatrix} 2/3 \\ 1/(3\sqrt{3}) \end{pmatrix},$$

$$\mathbf{f}_3(\mathbf{x}_4) = \begin{pmatrix} 11/18 \\ 1/(6\sqrt{3}) \end{pmatrix}, \quad \mathbf{f}_3(\mathbf{x}_5) = \begin{pmatrix} 2/3 \\ 0 \end{pmatrix}$$

$$\mathbf{f}_4(\mathbf{x}_1) = \begin{pmatrix} 2/3 \\ 0 \end{pmatrix}, \quad \mathbf{f}_4(\mathbf{x}_2) = \begin{pmatrix} 7/9 \\ 0 \end{pmatrix}, \quad \mathbf{f}_4(\mathbf{x}_3) = \begin{pmatrix} 5/6 \\ 1/(6\sqrt{3}) \end{pmatrix},$$

$$\mathbf{f}_4(\mathbf{x}_4) = \begin{pmatrix} 8/9 \\ 0 \end{pmatrix}, \quad \mathbf{f}_4(\mathbf{x}_5) = \begin{pmatrix} 1 \\ 0 \end{pmatrix}.$$

Thus there are 20 points (vectors), but three, of course, are the same

$$\mathbf{f}_3(\mathbf{x}_5) = \mathbf{f}_4(\mathbf{x}_1) = \begin{pmatrix} 2/3 \\ 0 \end{pmatrix},$$

$$\mathbf{f}_1(\mathbf{x}_5) = \mathbf{f}_2(\mathbf{x}_1) = \begin{pmatrix} 1/3 \\ 0 \end{pmatrix},$$

$$\mathbf{f}_3(\mathbf{x}_1) = \mathbf{f}_2(\mathbf{x}_5) = \begin{pmatrix} 1/2 \\ 1/(2\sqrt{3}) \end{pmatrix}.$$

These three points we have already from the first step. Note that

$$\mathbf{f}_1(\mathbf{x}_1) = \mathbf{x}_1, \qquad \mathbf{f}_5(\mathbf{x}_5) = \mathbf{x}_5.$$

Problem 37. Consider the Euclidean space $\mathbb{R}^3$. The *Menger sponge* is constructed as follows. The starting point is the unit cube $[0,1]^3 \equiv [0,1] \times [0,1] \times [0,1]$ with 8 vertices, 12 edges and volume 1. One subdivides this cube into $27 = 3^3$ smaller equal cubes by trisecting the edges (which have length 1). Thus the scaling factor is $1/3$. The trema to remove consists of the center cube and the 6 cubes in the centers of the faces of the cube. Hence 20 cubes remain each with volume $1/27 = 1/3 \cdot 1/3 \cdot 1/3$. The boundary of these 20 cubes must also remain in order for the set to be compact. Next we apply the same approach to the remaining 20 cubes and obtain $20^2 = 400$ cubes each with length $1/9$ and volume $1/9 \cdot 1/9 \cdot 1/9$. The Menger sponge is the set of points which remain if one applies this process infinitely often. Find the fractal dimension of the Menger sponge. Each face of the Menger sponge is a *Sierpinski carpet*.

Solution 37. At the n-th step we have $N_n = 20^n$ cubes each of length $L_n = (1/3)^n$. The fractal dimension follows as

$$D = -\lim_{n\to\infty} \frac{\ln(N_n)}{\ln(L_n)} = -\lim_{n\to\infty} \frac{\ln(20^n)}{\ln((1/3)^n)} = \frac{\ln(20)}{\ln(3)} = 2.726833...$$

Problem 38. Consider the Euclidean space $\mathbb{R}^3$ and the unit cube $[0,1]^3$. The *Menger sponge* is constructed by the 20 maps ($j = 1, \ldots, 20$)

$$\mathbf{f}_j(x_1, x_2, x_3) = \frac{1}{3} \begin{pmatrix} 1 & 0 & 0 \\ 0 & 1 & 0 \\ 0 & 0 & 1 \end{pmatrix} \begin{pmatrix} x_1 \\ x_2 \\ x_3 \end{pmatrix} + \mathbf{t}_j$$

where $\mathbf{t}_j = (\, t_{j,1} \quad t_{j,2} \quad t_{j,3}\,)^T$ for $t_{j,1}, t_{j,2}, t_{j,3} \in \{0, 1/3, 2/3\}$, except for the six case when exactly two coordinates are $1/3$ and the case when all three coordinates are $1/3$. This leads to the cases

$$\begin{pmatrix} 0 \\ 0 \\ 0 \end{pmatrix}, \begin{pmatrix} 0 \\ 0 \\ 1/3 \end{pmatrix}, \begin{pmatrix} 0 \\ 0 \\ 2/3 \end{pmatrix}, \begin{pmatrix} 0 \\ 1/3 \\ 0 \end{pmatrix}, \begin{pmatrix} 0 \\ 1/3 \\ 2/3 \end{pmatrix},$$

$$\begin{pmatrix} 0 \\ 2/3 \\ 0 \end{pmatrix}, \quad \begin{pmatrix} 0 \\ 2/3 \\ 1/3 \end{pmatrix}, \quad \begin{pmatrix} 0 \\ 2/3 \\ 2/3 \end{pmatrix}, \quad \begin{pmatrix} 1/3 \\ 0 \\ 0 \end{pmatrix}, \quad \begin{pmatrix} 1/3 \\ 0 \\ 2/3 \end{pmatrix},$$

$$\begin{pmatrix} 1/3 \\ 2/3 \\ 0 \end{pmatrix}, \quad \begin{pmatrix} 1/3 \\ 2/3 \\ 2/3 \end{pmatrix}, \quad \begin{pmatrix} 2/3 \\ 0 \\ 0 \end{pmatrix}, \quad \begin{pmatrix} 2/3 \\ 0 \\ 1/3 \end{pmatrix}, \quad \begin{pmatrix} 2/3 \\ 0 \\ 2/3 \end{pmatrix},$$

$$\begin{pmatrix} 2/3 \\ 1/3 \\ 0 \end{pmatrix}, \quad \begin{pmatrix} 2/3 \\ 1/3 \\ 2/3 \end{pmatrix}, \quad \begin{pmatrix} 2/3 \\ 2/3 \\ 0 \end{pmatrix}, \quad \begin{pmatrix} 2/3 \\ 2/3 \\ 1/3 \end{pmatrix}, \quad \begin{pmatrix} 2/3 \\ 2/3 \\ 2/3 \end{pmatrix}.$$

Consider $\mathbf{f}_{20}$

$$\mathbf{f}_{20}(\mathbf{x}) = \frac{1}{3}\begin{pmatrix} 1 & 0 & 0 \\ 0 & 1 & 0 \\ 0 & 0 & 1 \end{pmatrix}\begin{pmatrix} x_1 \\ x_2 \\ x_3 \end{pmatrix} + \begin{pmatrix} 2/3 \\ 2/3 \\ 2/3 \end{pmatrix}.$$

Apply it to the vertices

$$\begin{pmatrix} 0 \\ 0 \\ 0 \end{pmatrix}, \quad \begin{pmatrix} 1 \\ 0 \\ 1 \end{pmatrix}, \quad \begin{pmatrix} 0 \\ 1 \\ 0 \end{pmatrix}.$$

Solution 38. For the first vertex we obtain

$$\begin{pmatrix} 0 \\ 0 \\ 0 \end{pmatrix} \mapsto \begin{pmatrix} 2/3 \\ 2/3 \\ 2/3 \end{pmatrix} \mapsto \begin{pmatrix} 8/9 \\ 8/9 \\ 8/9 \end{pmatrix}.$$

The vectors tend to $(1\ 1\ 1)^T$. For the second vertex we obtain

$$\begin{pmatrix} 1 \\ 0 \\ 1 \end{pmatrix} \mapsto \begin{pmatrix} 1 \\ 2/3 \\ 1 \end{pmatrix} \mapsto \begin{pmatrix} 1 \\ 8/9 \\ 1 \end{pmatrix}.$$

The vectors tend to $(1\ 1\ 1)^T$. For the third vertex we obtain

$$\begin{pmatrix} 0 \\ 1 \\ 0 \end{pmatrix} \mapsto \begin{pmatrix} 2/3 \\ 1 \\ 2/3 \end{pmatrix} \mapsto \begin{pmatrix} 8/9 \\ 1 \\ 8/9 \end{pmatrix}.$$

The vectors tend to $(1\ 1\ 1)^T$.

Problem 39. The complex plane and the Riemann sphere (or extended complex plane) is denoted by $\mathbb{C}$ and $\hat{\mathbb{C}} = \mathbb{C} \cup \infty$, respectively. For any complex-valued rational function f on the Riemann sphere $\hat{\mathbb{C}}$ such that the point ∞ is an attracting fixed point of f, one defines

$$\mathcal{L}(f) = \text{the basin attraction of } \infty \text{ for the map} f$$
$$\mathcal{K}(f) = \hat{\mathbb{C}} \setminus \mathcal{L}(f)$$
$$\mathcal{J}(f) = \partial(\mathbb{K}(f))$$

where $\partial(S)$ denotes the boundary of a set S. If the function f is a polynomial, then $\mathcal{J}(f)$ is the Julia set of f, $\mathcal{K}(f)$ is the filled Julia set with $\mathcal{K}(f)$ given by

$$\mathcal{K}(f) = \{\, z \in \hat{\mathbb{C}} : f^{(n)}(z) \nrightarrow \infty \text{ as } n \to \infty \}.$$

Find the filled Julia set and the Julia set for

$$f(z) = z^2.$$

Solution 39. All points in the complex plane either tend to the fixed point 0 if $|z_0| < 1$ or

$$\lim_{n \to \infty} f^{(n)}(z_0) = \infty$$

if $|z_0| > 1$ or $|f^{(n)}(z_0)| = 1$ if $|z_0| = 1$. Thus the unit circle $|z| = 1$ in the complex plane is the boundary between the two basins of attraction (i.e. 0 and ∞). This boundary set is the Julia set. Thus the filled Julia set $\mathcal{K}(f)$ is the set $\{z : |z| \leq 1\}$.

Problem 40. Consider the map

$$f_c(z) = z^2 + c$$

with $c = 1/4$. The only fixed point is $z^* = 1/2$.
(i) Is $z = 1/2$ an element of the filled Julia set?
(ii) Is $z = -1/2$ an element of the filled Julia set?
(iii) Is $z = i/2$ an element of the filled Julia set?
(iv) Is $z = -i/2$ an element of the filled Julia set?

Solution 40. (i) We have

$$f_{1/4}(1/2) = \frac{1}{2}$$

i.e. we have a fixed point and $z = 1/2$ is an element of the filled Julia set.
(ii) We have

$$f_{1/4}(-1/2) = \frac{1}{2}$$

i.e. the orbit tends to the fixed point $1/2$. Thus $-1/2$ is an element of the filled Julia set.
(iii) We have

$$f_{1/4}(i/2) = 0, \quad f_{1/4}(0) = \frac{1}{4}, \quad f_{1/4}(1/4) = \frac{5}{16}.$$

The orbit tends to the fixed point $1/2$, i.e. $i/2$ is an element of the filled Julia set.
(iv) We have $f_{1/4}(-i/2) = 0$. Thus we ave the case (iii).

Problem 41. The *Riemann-Liouville definition* for the *fractional derivative* of a function f is given by

$$\frac{d^\alpha f(t)}{dt^\alpha} = \frac{1}{\Gamma(n-\alpha)} \frac{d^n}{dt^n} \int_{\tau=0}^{\tau=t} \frac{f(\tau)}{(t-\tau)^{\alpha-n+1}} d\tau$$

where $\Gamma(.)$ is the gamma function and the integer n is given by $n-1 \leq \alpha < n$. Let $f(t) = t^2$. Find the fractional derivative of f with $\alpha = 1/2$.

Solution 41. Since $\alpha = 1/2$ we have $n = 1$ and $\alpha - n + 1 = 1/2$. Now we have

$$\int_{\tau=0}^{\tau=t} \frac{\tau^2}{\sqrt{t-\tau}} d\tau = -\frac{2(3\tau^2 + 4t\tau + 8t^2)}{15} \sqrt{t-\tau} \Big|_{\tau=0}^{\tau=t} = \frac{16}{15} t^{5/2}.$$

Since $dt^{5/2}/dt = 5t^{3/2}/2$ and $\Gamma(1/2) = \sqrt{\pi}$ we obtain

$$\frac{d^{1/2} f(t)}{dt^{1/2}} = \frac{8}{3\sqrt{\pi}} t^{3/2}.$$

Problem 42. Consider the equilateral triangle with the vertices

$$\mathbf{x}_1 = \begin{pmatrix} 0 \\ 0 \end{pmatrix}, \quad \mathbf{x}_2 = \begin{pmatrix} 1 \\ 0 \end{pmatrix}, \quad \mathbf{x}_3 = \begin{pmatrix} 1/2 \\ \sqrt{3}/2 \end{pmatrix}.$$

(i) Find the area of the triangle.
(ii) Let $\mathbf{x} = \begin{pmatrix} x_1 \\ x_2 \end{pmatrix} \in \mathbb{R}^2$. Consider the three contracting maps

$$f_1(\mathbf{x}) = \frac{1}{2}\mathbf{x}, \quad f_2(\mathbf{x}) = \frac{1}{2}\mathbf{x} + \begin{pmatrix} 1/2 \\ 0 \end{pmatrix}, \quad f_3(\mathbf{x}) = \frac{1}{2}\mathbf{x} + \begin{pmatrix} 1/4 \\ \sqrt{3}/4 \end{pmatrix}.$$

Find

$$f_1(\mathbf{x}_1), \quad f_2(\mathbf{x}_1), \quad f_3(\mathbf{x}_1)$$

$$f_1(\mathbf{x}_2), \quad f_2(\mathbf{x}_2), \quad f_3(\mathbf{x}_2)$$
$$f_1(\mathbf{x}_3), \quad f_2(\mathbf{x}_3), \quad f_3(\mathbf{x}_3).$$

Show that we obtain the vertices given above and three new vertices that describe an inscribed equilateral triangle. Find the area of this triangle.

Solution 42. (i) The area is given by

$$A = \pm\frac{1}{2}\det\begin{pmatrix} x_{11} & x_{12} & 1 \\ x_{21} & x_{22} & 1 \\ x_{31} & x_{32} & 1 \end{pmatrix} = \pm\frac{1}{2}\det\begin{pmatrix} 0 & 0 & 1 \\ 1 & 0 & 1 \\ 1/2 & \sqrt{3}/2 & 1 \end{pmatrix} = \frac{\sqrt{3}}{4}.$$

(ii) We obtain

$$f_1(\mathbf{x}_1) = \begin{pmatrix} 0 \\ 0 \end{pmatrix}, \quad f_2(\mathbf{x}_1) = \begin{pmatrix} 1/2 \\ 0 \end{pmatrix}, \quad f_3(\mathbf{x}_1) = \begin{pmatrix} 1/4 \\ \sqrt{3}/4 \end{pmatrix}$$

$$f_1(\mathbf{x}_2) = \begin{pmatrix} 1/2 \\ 0 \end{pmatrix}, f_2(\mathbf{x}_2) = \begin{pmatrix} 1 \\ 0 \end{pmatrix}, \quad f_3(\mathbf{x}_2) = \begin{pmatrix} 3/4 \\ \sqrt{3}/4 \end{pmatrix}$$

$$f_1(\mathbf{x}_3) = \begin{pmatrix} 1/4 \\ \sqrt{3}/4 \end{pmatrix}, \quad f_2(\mathbf{x}_3) = \begin{pmatrix} 3/4 \\ \sqrt{3}/4 \end{pmatrix}, \quad f_3(\mathbf{x}_3) = \begin{pmatrix} 1/2 \\ \sqrt{3}/2 \end{pmatrix}.$$

Hence we obtain the vertices given above and the three new vertices

$$\begin{pmatrix} 1/2 \\ 0 \end{pmatrix}, \quad \begin{pmatrix} 1/4 \\ \sqrt{3}/4 \end{pmatrix}, \quad \begin{pmatrix} 3/4 \\ \sqrt{3}/4 \end{pmatrix}.$$

The area of the triangle with these vertices is $\sqrt{3}/16$. Thus the area this triangle is four times smaller than the area of the original triangle. Repeating this process we obtain the *Sierpinski triangle.*

Problem 43. Let $0 < r < 1$ and $\phi \in \mathbb{R}$. Consider the two maps

$$f_0(z) = re^{i\phi}z, \qquad f_1(z) = re^{i\phi} + (1 - re^{i\phi})z.$$

Find the fixed points of the two maps f_0 and f_1. Show that $f_0(1) = f_1(0)$.

Solution 43. From $f_0(z^*) = z^*$ we obtain the fixed point $z^* = 0$. From $f_1(z^*) = z^*$ we obtain the fixed point $z^* = 1$. We have $f_0(1) = re^{i\phi}$ and $f_1(0) = re^{i\phi}$.

Problem 44. Let $d = 2, 3, \ldots$. Consider the function $f : \mathbb{C} \to \mathbb{C}$

$$f(z) = z^d.$$

Find $f^{(n)}(z)$ and $\lim_{n\to\infty} f^{(n)}(z)$.

Solution 44. We obtain

$$f^{(n)} = z^{d^n}$$

and

$$\lim_{n\to\infty} f^{(n)}(z) = \begin{cases} 0 \text{ if } |z| < 1 \\ \infty \text{ if } |z| > 1. \end{cases}$$

Problem 45. Let $\mathbb{C}$ be the complex plane. Let $c \in \mathbb{C}$. The Mandelbrot set M is defined as follows

$$M := \{ c \in \mathbb{C} : c, c^2 + c, (c^2 + c)^2 + c, \ldots \not\to \infty \}. \qquad (1)$$

(i) Show that to find the Mandelbrot set one has to study the recursion relation

$$z_{t+1} = z_t^2 + c \qquad (2)$$

where $t = 0, 1, 2, \ldots$ and $z_0 = 0$.
(ii) Write the recursion relation in real and imaginary part. For a given $c \in \mathbb{C}$ (or $(c_1, c_2) \in \mathbb{R}^2$) we can now study whether or not c belongs to M.
(iii) Show that $(c_1, c_2) = (0, 0)$ belongs to M.

Solution 45. (i) This is obvious since

$$z_1 = c, \;\; z_2 = c^2 + c, \;\; z_3 = (c^2 + c)^2 + c \qquad (3)$$

etc..
(ii) Since $z = x + iy$ and $c = c_1 + ic_2$ with $x, y, c_1, c_2 \in \mathbb{R}$ we can write (2) as

$$x_{t+1} = x_t^2 - y_t^2 + c_1, \qquad y_{t+1} = 2x_t y_t + c_2.$$

(iii) With the initial value $(x_0, y_0) = (0, 0)$ we obtain . If $|z| > 2$, then

$$|z^2 + c| \geq |z^2| - |c| > 2|z| - |c|.$$

If $|z| \geq |c|$, then $2|z| - |c| > |z|$. So, if $|z| > 2$ and $|z| \geq c$, $|z^2 + c| > |z|$, so that the sequence is increasing. It takes more work to prove it is unbounded and diverges.

Problem 46. Let $c \in \mathbb{C}$. Consider the function

$$f_c(z) = z^2 + c.$$

Show that the Mandelbrot set M is contained in the disk of radius 2 in the complex plane, i.e.

$$M \subset \{ c \in \mathbb{C} : |c| \leq 2 \}.$$

Solution 46. Let $|c| > 2$ and $z_n := f_c^{(n)}(0)$. Then

$$|z_{n+1}| \geq |z_n^2| - |c| = |z_n|(|z_n| - 1) + (|z_n| - |c|).$$

Since $z_1 = |c| > 2$ by assumption we find that $|z_n|$ is an increasing sequence. Hence

$$|z_{n+1}| \geq |z_1|(|z_1| - 1)^n = |c|(|c| - 1)^n.$$

Consequently $|z_n| \to \infty$ and c is not in the Mandelbrot set.

Problem 47. Let $c \in \mathbb{C}$. Consider the function $f : \mathbb{C} \to \mathbb{C}$

$$f_c(z) = z^2 + c.$$

A point c is called a *Misiurewicz point* if 0 is strictly preperiodic for f_c. One calls $c \in \mathbb{C}$ a Misiurewicz point of type (m, n) if $m \geq 1$ is the smallest integer such that $f_c^{(n)}(0)$ is periodic and n is the primitive period of $f_c^{(m)}(0)$. Give examples for Misiurewicz points.

Solution 47. A Misiurewicz point is $c = i$. One has

$$0, \; f_i(0) = i, \; f_i(i) = -1 + i, \; f_i(-1 + i) = -i, \; f_i(-i) = -1 + i.$$

Another Misiurewicz point is $c = -2$. One has

$$0, \; f_{-2}(0) = -2, \; f_{-2}(-2) = 2, \; f_{-2}(2) = 2.$$

Problem 48. Consider the logistic map $f_r : \mathbb{R} \to \mathbb{R}$ given by

$$f_r(x) = rx(1 - x) \tag{1}$$

where $r > 2 + \sqrt{5}$. Let us restrict the map to $f_r : \Lambda \to \Lambda$ (Cantor set). Consider the *shift map* $\sigma : \Sigma \to \Sigma$ given by

$$\sigma(s_0 s_1 s_2 \ldots) := s_1 s_2 s_3 \ldots \tag{2}$$

In other words, the shift map forgets the first digit of the sequence. Construct a topological conjugacy from f_r to σ.

Solution 48. First we define the function $\psi : \Lambda \to \Sigma$. We set

$$I_0 = \left[0, \frac{1}{2} - \frac{\sqrt{r^2 - 4r}}{2r}\right], \qquad I_1 = \left[\frac{1}{2}, \frac{\sqrt{r^2 - 4r}}{2r}, 1\right]. \tag{3}$$

Now we have

$$\Lambda_1 := \{x : f_r(x) \in [0, 1]\} = I_0 \cup I_1. \tag{4}$$

Since Λ is a subset of Λ_1, Λ is also included into $I_0 \cup I_1$. For each x in Λ, define the sequence

$$\psi(x) = s_0 s_1 s_2 \ldots \tag{5}$$

in Σ so that $f^{(n)}(x)$ is in I_{s_n} for each n. The digit in s_n in $\psi(x) = s_0 s_1 s_2 \ldots$ is 0 if and only if $f_r^{(n)}(x)$ is in I_0 and $s_n = 1$ if and only if $f_r^{(n)}(x)$ is in I_1. Thus ψ is well-defined. It is straightforward to show that (i) ψ is one to one and onto, (ii) ψ is continuous, (iii) ψ^{-1} is continuous, (iv) $\psi \circ f_r = \sigma \circ \psi$.

Problem 49. Let $\mathbb{C}$ be the complex plane. Let $c \in \mathbb{C}$. The *Mandelbrot set* M is defined as

$$M := \{\, c \in \mathbb{C} \,:\, c,\, c^2 + c,\, (c^2 + c)^2 + c, \ldots, \not\to \infty \,\}.$$

To find the Mandelbrot set we study the recursion relation

$$z_{t+1} = z_t^2 + c, \qquad t = 0, 1, 2, \ldots$$

with the initial value $z_0 = 0$ and whether z_t escapes to infinity. For example $c = 0$ and $c = 1/4 + i/4$ belong to the Mandelbrot set. The point $c = 1/2$ does not belong to the Mandelbrot set. Write a C++ program using the complex class of STL to find the Mandelbrot set. The output should be written to a file `Mandel.pnm` (portable anymap utilities). This file can then be used to display the fractal.

Solution 49. Using `complex<double> z` we have the implementation

```
// mandelbrot.cpp

#include <complex>
#include <fstream>
#include <iostream>
using namespace std;

int mandeltest(const complex<double> &c,int maxiter)
{
  int j;
  complex<double> z;
  for(j=0,z=0.0;j<maxiter;j++)
  {
  z=z*z+c;
  if(abs(z)>100) return j/int(1.0+exp(-abs(z)));
  }
  return j;
}

int main(void)
```

```
{
  int maxiter=255, dpi=180, xcm=100, ycm=200;
  int i, j;
  double minx=-1.33, maxx=-1.25, miny=-0.20, maxy;
  double x, y, stepx, stepy;
  int xcount=int(dpi*xcm/2.5), ycount=int(dpi*ycm/2.5);
  ofstream mandel("mandel.pnm");
  maxy=(maxx-minx)*ycount/xcount+miny;
  stepx=(maxx-minx)/xcount;
  stepy=(maxy-miny)/ycount;
  cout << xcount << " x " << ycount << endl;
  mandel << "P6 " << xcount << " "
         << ycount << " " << maxiter << endl;
  for(i=0,y=miny;i<ycount;y+=stepy,i++)
  {
  cout.precision(3);
  cout << 100.0*i/ycount << "% \r";
  cout.flush();
  for(j=0,x=minx;j<xcount;x+=stepx,j++)
  {
  int k=mandeltest(complex<double>(x,y),maxiter);
  unsigned char c;
  c=30+char(205*((x-minx)*k/((maxx-minx)*maxiter)));
  mandel.write((char*)&c,1);
  c=char(255*((y-miny)*k/((maxy-miny)*maxiter)));
  c=20;
  mandel.write((char*)&c,1);
  c=30+char(225*double(k)/maxiter);
  mandel.write((char*)&c,1);
  }
}
  mandel.close();
  cout << endl;
  return 0;
}
```

3.3 Supplementary Problems

Problem 1. Study the fractal generated by the 3×3 matrix

$$M = \begin{pmatrix} 1 & 1 & 1 \\ 0 & 1 & 1 \\ 1 & 1 & 1 \end{pmatrix}$$

and the Kronecker product $\otimes$. Note that

$$M \otimes M = \begin{pmatrix} 1 & 1 & 1 & 1 & 1 & 1 & 1 & 1 & 1 \\ 0 & 1 & 1 & 0 & 1 & 1 & 0 & 1 & 1 \\ 1 & 1 & 1 & 1 & 1 & 1 & 1 & 1 & 1 \\ 0 & 0 & 0 & 1 & 1 & 1 & 1 & 1 & 1 \\ 0 & 0 & 0 & 0 & 1 & 1 & 0 & 1 & 1 \\ 0 & 0 & 0 & 1 & 1 & 1 & 1 & 1 & 1 \\ 1 & 1 & 1 & 1 & 1 & 1 & 1 & 1 & 1 \\ 0 & 1 & 1 & 0 & 1 & 1 & 0 & 1 & 1 \\ 1 & 1 & 1 & 1 & 1 & 1 & 1 & 1 & 1 \end{pmatrix}.$$

Problem 2. Consider the 3×3 binary matrix

$$A = \begin{pmatrix} 1 & 0 & 1 \\ 0 & 1 & 0 \\ 1 & 0 & 1 \end{pmatrix}$$

where 1 is identified with a black pixel and 0 with a white pixel. What fractal is generated by

$$A, \quad A \otimes A, \quad A \otimes A \otimes A, \ \ldots$$

Give the fractal dimension. Note that

$$A \otimes A = \begin{pmatrix} 1 & 0 & 1 & 0 & 0 & 0 & 1 & 0 & 1 \\ 0 & 1 & 0 & 0 & 0 & 0 & 0 & 1 & 0 \\ 1 & 0 & 1 & 0 & 0 & 0 & 1 & 0 & 1 \\ 0 & 0 & 0 & 1 & 0 & 1 & 0 & 0 & 0 \\ 0 & 0 & 0 & 0 & 1 & 0 & 0 & 0 & 0 \\ 0 & 0 & 0 & 1 & 1 & 0 & 0 & 0 & 0 \\ 1 & 0 & 1 & 0 & 0 & 0 & 1 & 0 & 1 \\ 0 & 1 & 0 & 0 & 0 & 0 & 0 & 1 & 0 \\ 1 & 0 & 1 & 0 & 0 & 0 & 1 & 0 & 1 \end{pmatrix}.$$

Problem 3. What fractal is generated by the 4×4 matrix

$$M = \begin{pmatrix} 1 & 0 \\ 1 & 1 \end{pmatrix} \otimes \begin{pmatrix} 1 & 1 \\ 0 & 1 \end{pmatrix} \equiv \begin{pmatrix} 1 & 1 & 0 & 0 \\ 0 & 1 & 0 & 0 \\ 1 & 1 & 1 & 1 \\ 0 & 1 & 0 & 1 \end{pmatrix}.$$

and the Kronecker product? The two 2×2 matrices on the right-hand side are elements of the Lie group $SL(2, \mathbb{R})$.

Problem 4. Gray scale fractal images can be produced if the entries of the matrix are allowed to take on values between 0 and 1. Find the gray scale fractal generated by the 3×3 matrix

$$M = \begin{pmatrix} 1.0 & 0.5 & 1.0 \\ 0.5 & 1.0 & 0.5 \\ 1.0 & 0.5 & 1.0 \end{pmatrix}$$

and the Kronecker product.

Problem 5. Consider the unit square $[0,1]^2$ with the four vertices $(0,0)$, $(0,1)$, $(1,0)$, $(1,1)$. Study the four maps

$$\mathbf{f}_1 = \frac{1}{2}\begin{pmatrix} 0 & 1 \\ 1 & 0 \end{pmatrix}\begin{pmatrix} x_1 \\ x_2 \end{pmatrix}, \qquad \mathbf{f}_2 = \frac{1}{2}\begin{pmatrix} 0 & 1 \\ 1 & 0 \end{pmatrix}\begin{pmatrix} x_1 \\ x_2 \end{pmatrix} + \begin{pmatrix} 1/2 \\ 0 \end{pmatrix},$$

$$\mathbf{f}_3 = \frac{1}{2}\begin{pmatrix} 0 & 1 \\ 1 & 0 \end{pmatrix}\begin{pmatrix} x_1 \\ x_2 \end{pmatrix} + \begin{pmatrix} 0 \\ 1/2 \end{pmatrix}, \qquad \mathbf{f}_4 = \frac{1}{2}\begin{pmatrix} 0 & 1 \\ 1 & 0 \end{pmatrix}\begin{pmatrix} x_1 \\ x_2 \end{pmatrix} + \begin{pmatrix} 1/2 \\ 1/2 \end{pmatrix}.$$

Problem 6. Study the iterated functions system

$$\mathbf{f}_1(x_1, x_2) = \frac{1}{2}\begin{pmatrix} 1 & 0 \\ 0 & 1 \end{pmatrix}\begin{pmatrix} x_1 \\ x_2 \end{pmatrix}$$

$$\mathbf{f}_2(x_1, x_2) = \frac{1}{2}\begin{pmatrix} \cos(\theta) & -\sin(\theta) \\ \sin(\theta) & \cos(\theta) \end{pmatrix}\begin{pmatrix} x_1 \\ x_2 \end{pmatrix} + \begin{pmatrix} 1/3 \\ 1/3 \end{pmatrix}.$$

Apply it to the vertices $(0,0)$, $(0,1)$, $(1,0)$, $(1,1)$ of the unit square $[0,1]^2$.

Problem 7. Study the pair of maps

$$f_1(z) = \frac{1}{1+z}, \qquad f_2(z) = \frac{z}{1+z}.$$

Problem 8. The *Appolonian gasket* is a subset of the Euclidean space $\mathbb{R}^2$. Find the fractal dimension of the Appolonian gasket.

Problem 9. Let $\alpha > 0$. Show that the box counting dimension of the countable set

$$\left\{ \frac{1}{n^\alpha} : n \in \mathbb{N} \right\}$$

is equal to $1/(\alpha+1)$. For the box size we choose

$$\varepsilon := \frac{1}{n^\alpha} - \frac{1}{(n+1)^\alpha} \quad \Rightarrow \quad \varepsilon = \frac{(n+1)^\alpha - n^\alpha}{n^\alpha (n+1)^\alpha}.$$

Problem 10. Consider the iterated function system (*Sierpinski triangle*)

$$\mathbf{f}_1(x_1, x_2) = \frac{1}{2}\begin{pmatrix} 1 & 0 \\ 0 & 1 \end{pmatrix}\begin{pmatrix} x_1 \\ x_2 \end{pmatrix}$$

$$\mathbf{f}_2(x_1, x_2) = \frac{1}{2}\begin{pmatrix} 1 & 0 \\ 0 & 1 \end{pmatrix}\begin{pmatrix} x_1 \\ x_2 \end{pmatrix} + \begin{pmatrix} 1/2 \\ 0 \end{pmatrix}$$

$$\mathbf{f}_3(x_1, x_2) = \frac{1}{2}\begin{pmatrix} 1 & 0 \\ 0 & 1 \end{pmatrix}\begin{pmatrix} x_1 \\ x_2 \end{pmatrix} + \begin{pmatrix} 0 \\ 1/2 \end{pmatrix}.$$

Apply the maps to the vertices of the unit square $[0, 1]^2$. Discuss.

Problem 11. Starting from an equilateral triangle the first two steps in construction of the triangular *Sierpinski gasket* are given in the figure. Find the box counting dimension (capacity).

Problem 12. The *Sierpinski gasket* in d dimensional Euclidean space $(d \geq 2)$ is constructed as follows: One starts with a d-dimensional hypertetrahedron. The midpoints of the edges are connected, creating $(d+1)$ smaller hypertetrahedra. The set at the centre (bounded by the faces of these new tetrahedra) is then removed. This procedure is repeated for the $(d+1)$ new tetrahedra and so forth. Show that the capacity is given by

$$C = \frac{\ln(d+1)}{\ln(2)}.$$

Problem 13. Consider the 3×3 binary matrix

$$B = \begin{pmatrix} 1 & 0 & 1 \\ 0 & 0 & 0 \\ 1 & 0 & 1 \end{pmatrix}.$$

What fractal is generated by $B \otimes B$, $B \otimes B \otimes B$, etc. Note that

$$B \otimes B = \begin{pmatrix} 1 & 0 & 1 & 0 & 0 & 0 & 1 & 0 & 1 \\ 0 & 0 & 0 & 0 & 0 & 0 & 0 & 0 & 0 \\ 1 & 0 & 1 & 0 & 0 & 0 & 1 & 0 & 1 \\ 0 & 0 & 0 & 0 & 0 & 0 & 0 & 0 & 0 \\ 0 & 0 & 0 & 0 & 0 & 0 & 0 & 0 & 0 \\ 0 & 0 & 0 & 0 & 0 & 0 & 0 & 0 & 0 \\ 1 & 0 & 1 & 0 & 0 & 0 & 1 & 0 & 1 \\ 0 & 0 & 0 & 0 & 0 & 0 & 0 & 0 & 0 \\ 1 & 0 & 1 & 0 & 0 & 0 & 1 & 0 & 1 \end{pmatrix}$$

Problem 14. Let $f_1, f_2 : \mathbb{R} \to \mathbb{R}$ be an iterated function system given by

$$f_1(x) = \frac{1}{4}x, \qquad f_2(x) = \frac{1}{2}x + \frac{1}{2}.$$

Describe the attractor of $\{f_1, f_2\}$ and find its Hausdorff dimension and capacity. Note that the fixed point of f_1 is $x^* = 0$ and the fixed point of f_2 is $x^* = 1$.

Problem 15. Consider the map $f : \mathbb{C} \to \mathbb{C}$

$$f(z) = z^2 + i.$$

Note that

$$f(1) = 1 + i, \quad f(-1) = 1 + i, \quad f(i) = -1 + i, \quad f(-i) = -1 + i.$$

(i) Find $f(f(1))$, $f(f(f(1)))$, ... Discuss.
(ii) Find $f(f(-1))$, $f(f(f(-1)))$, ... Discuss.
(iii) Find $f(f(i))$, $f(f(f(i)))$, ... Discuss.
(iv) Find $f(f(-i))$, $f(f(f(-i)))$, ... Discuss.
(v) Show that for both (iii) and (iv) we have a repelling cycle.

Problem 16. Let $x \in [0, 1]$. Consider the logistic map

$$f_r(x) = rx(1 - x)$$

with $r = r_\infty \approx 3.570....$ Show that the corresponding invariant set $A \subset [0, 1]$ has both Hausdorff and box counting dimensions approximatly equal to ≈ 0.538.

Problem 17. The construction of a two-dimensional Cantor set is as follows: We start from the unit square $[0, 1] \times [0, 1]$. In the first step we construct four subsets of the unit square given by

$$1)\ [0, 1/3] \times [0, 1/3], \ 2)\ [2/3, 1] \times [0, 1/3],$$

$$3)\ [0,1/3]\times[2/3,1],\quad 4)\ [2/3,1]\times[2/3,1].$$

Within each of these four subsets we construct again four subsets with the scaling factors 1/3. Thus for the first subset $[0,1/3]\times[0,1/3]$ we obtain the four subsets

$$[0,1/9]\times[0,1/9],\quad [2/9,1/3]\times[0,1/9],$$
$$[0,1/9]\times[2/9,1/3],\quad [2/9,1/3]\times[2/9,1/3].$$

Analogously we do the construction for the other three subsets (squares). We repeat this process now up to infinity. Find the Hausdorff dimension and the capacity of this set.

Problem 18. Study the four *similitudes*

$$\mathbf{f}_1(x_1,x_2)=\frac{3}{4}\begin{pmatrix}1/\sqrt{2} & 1/\sqrt{2}\\ 1/\sqrt{2} & -1/\sqrt{2}\end{pmatrix}\begin{pmatrix}x_1\\ x_2\end{pmatrix}$$
$$\mathbf{f}_2(x_1,x_2)=\frac{3}{4}\begin{pmatrix}1/\sqrt{2} & 1/\sqrt{2}\\ 1/\sqrt{2} & -1/\sqrt{2}\end{pmatrix}\begin{pmatrix}x_1\\ x_2\end{pmatrix}+\begin{pmatrix}1/4\\ 0\end{pmatrix}$$
$$\mathbf{f}_3(x_1,x_2)=\frac{3}{4}\begin{pmatrix}1/\sqrt{2} & 1/\sqrt{2}\\ 1/\sqrt{2} & -1/\sqrt{2}\end{pmatrix}\begin{pmatrix}x_1\\ x_2\end{pmatrix}+\begin{pmatrix}0\\ 1/4\end{pmatrix}$$
$$\mathbf{f}_4(x_1,x_2)=\frac{3}{4}\begin{pmatrix}1/\sqrt{2} & 1/\sqrt{2}\\ 1/\sqrt{2} & -1/\sqrt{2}\end{pmatrix}\begin{pmatrix}x_1\\ x_2\end{pmatrix}+\begin{pmatrix}1/4\\ 1/4\end{pmatrix}$$

for the unit square $[0,1]^2$ with the vertices $(0,0)$, $(0,1)$, $(1,0)$, $(1,1)$.

Problem 19. Is the Cantor function (devil staircase) continuous?

Problem 20. The *Weierstrass function* $f:\mathbb{R}\to\mathbb{R}$

$$f(x)=\sum_{n=1}^{\infty}\frac{1}{n^2}\sin(n^2x)$$

is continuous but nowhere differentiable. Find the derivative in the sense of generalized functions. The multifractal structure of the Weierstrass function can be analyzed using wavelets of Lusin.

Problem 21. Consider the *Mandelbrot set*

$$M:=\{\,c\in\mathbb{C}\ :\ \text{for all } n\geq 1,\ |f_c^{(n)}(0)|\leq 2\,\}\quad\text{where}\quad f_c(z)=z^2+c.$$

M is a subset of the complex plane.
(i) Is $c=1$ an element of the Mandelbrot set?
(ii) Is $c=-1$ an element of the Mandelbrot set?
(iii) Is $c=i$ an element of the Mandelbrot set?

(iv) Is $c = -i$ an element of the Mandelbrot set?

Problem 22. Consider the map

$$f_c(z) = z^2 + c$$

with $c = 1/4$. Is $z = 3/5$ an element of the filled Julia set.

Problem 23. Let $I = [0,1]$. Find the attractor for the iterated function system

$$f_1(x) = \frac{1}{2}x, \qquad f_2(x) = \frac{1}{4}x + \frac{3}{4}.$$

Problem 24. Let

$$X = \begin{pmatrix} 1 & 1 \\ 1 & 1 \end{pmatrix}, \qquad B = \begin{pmatrix} 0 & 1 \\ 1 & 0 \end{pmatrix}.$$

Show that a checkerboard can be generated via

$$\left(\bigotimes_{j=1}^{n} X \right) \otimes B.$$

Problem 25. Let $c, z \in \mathbb{C}$, $|c| < 1$ and $|1 - c| < 1$. These conditions are satisfied by

$$c = \frac{1}{2}(1 + i).$$

Study the maps

$$f_0(z) = cz, \qquad f_1(z) = c + (1 - c)z.$$

First find the fixed points for f_0 and f_1. Since $c \neq 0$ the only fixed point for f_0 is $z^* = 0$ and for f_1 the only fixed point is $z^* = 1$.

Problem 26. Let $x \in [0,1]$. Consider the linear functions

$$f_L(x) = \frac{x}{3}, \qquad f_R(x) = \frac{1}{3}(2 + x).$$

(i) Show that the fixed points of f_L is 0. Show that the fixed points of f_R is 0.
(ii) Let $x = 1$. Find $f_L(x)$, $f_L(f_L(x))$, $f_L(f_L(f_L(x)))$, $\ldots$.
(iii) Let $x = 1$. Find $f_R(x)$, $f_R(f_R(x))$, $f_R(f_R(f_R(x)))$, $\ldots$.

(iv) Let $x = 1/2$. Find $f_L(x)$, $f_L(f_L(x))$, $f_L(f_L(f_L(x)))$, ... and $f_R(x)$, $f_R(f_R(x))$, $f_R(f_R(f_R(x)))$, ...

Problem 27. Let s be a positive number. Study the *Julia set* of the rational map

$$f(z) = \left(\frac{z^2+s-1}{2z+s-2}\right)^2 .$$

If s is large the Julia set is a *Jordan circle*. As $s \to \infty$ it becomes larger and more circular.

Problem 28. Solve the equation

$$f(x) = \frac{3}{2}(f(x-1) + f(3x+1))$$

together with the condition

$$\int_{-\infty}^{\infty} f(x)dx = 1$$

in the sense of generalized functions. What is the connection with the Cantor set?

Bibliography

Books

Alligood K. T., Sauer T. D. and Yorke J. A.
Chaos: An introduction to dynamical systems, Springer, New York, 1997

Antimirov M. Ya., Kolyshkin A. A. and Vaillancourt R.
Complex Variables, Academic Press, 1998

Arnold V. I. and Avez A.
Ergodic Problems in Classical Mechanics, Addison-Wesley, 1989

Arrowsmith D. K. and Place C. M.
An Introduction to Dynamical Systems, Cambridge University Press, 1990

Arrowsmith D. K. and Place C. M.
Dynamical Systems: Differential equations, maps and chaotic behaviour, Chapman and Hall, 1992

Baker G. L. and Gollub J. P.
Chaotic Dynamics: An Introduction, Cambridge University Press, 1990

Barnsley M. F.
Fractals Everywhere, Academic Press, Boston, 1988

Beardon A. F.
Iteration of Rational Functions, Springer Verlag, 1991

Beck C. and Schlögl F.
Thermodynamics of Chaotic Systems, Cambridge University Press, 1993

Berge P., Pomeau Y. and Vidal C.
Order within Chaos, Wiley, New York, 1984

Birkhoff G. D.
Collected Mathematical Papers, Vols. 1,2,3 Am. Math. Soc., Providence, 1950

Bowen R.
Methods of Symbolic Dynamics, Lecture Notes in Mathematics **470**, Springer Verlag, 1975

Bowen R.
Equilibrium States and the Ergodic Theory of Anosov Diffeomorphisms, Lecture Notes in Mathematics **470**, Springer Verlag, Berlin, 1975

Carleson L. and Gamelin T. G. W.
Complex Dynamics, Springer Verlag, 1993

Cvitanovic P.
Universality in Chaos, second edition, Adam Hilger, 1989

Choe Geon Ho
Computational Ergodic Theory, Algorithms and Computation in Mathematics, Vol. 13, Springer Verlag, 2005

Chow S. N. and Hale J. K.
Methods of Bifurcation Theory, Springer-Verlag, 1982

Collet P. and Eckmann J.-P.
Iterated Maps on the Intervals as Dynamical Systems, Birckhäuser, Boston, 1980

Devaney R. L.
An Introduction to Chaotic Dynamical Systems, second edition, Addison-Wesley, 1989

Edgar G. A.
Measure, Topology and Fractal Geometry, Springer, New York, 1990

Edgar G. A.
Integral, probability and fractal measures, Springer Verlag, 1998

Elaydi S. N.
An Introduction to Difference Equations, second edition, Springer, 1999

Falconer K.
Techniques in Fractal Geometry, John Wiley, Chichester, 1997

Feder J.
Fractals, Plenum Press, New York, 1988

Fröberg C. E.
Numerical Mathematics: Theory and Computer Applications, Benjamin-Cummings, 1985

Guckenheimer and Holmes P.
Nonlinear Oscillations, Dynamical Systems, and Bifurcations of Vector Fields, Springer Verlag, New York, 1983

Gulick D.
Encounters with Chaos and Fractals, second edition, CRC Press, 2012

Gutzwiller M. C.
Chaos in Classical and Quantum Mechanics, Springer Verlag, 1990

Holmgren R. A.
A First Course in Discrete Dynamical Systems, second edition. Springer, New York, 1996

Katok A. and Hasselblatt B.
Introduction to the Modern Theory of Dynamical Systems, Cambridge University Press, Cambridge, 1995

Krause U. and Nesemann T.
Differenzengleichungen und diskrete dynamische Systeme, Teubner, Stuttgart-Leipzig, 1999

Lang S.
Complex Analysis, second edition, Springer, New York, 1985

Lynch S.
Dynamical Systems with Applications Using Maple, second edition, Springer, New York, 2014

Mandelbrot B.
The Fractal Geometry of Nature, W. H. Freeman, 1983

Menger K.
Dimensionstheorie, B. G. Teubner, 1928

Nakamura K.
Quantum Chaos - A New Paradigm of Nonlinear Dynamics, Cambridge University Press, Cambridge, 1993

Nitecki Z.
Differentiable dynamics, Cambridge, MA, London: MIT Press, 1971

Ott E.
Chaos in Dynamical Systems, Cambridge University Press, 1993

Palis J. and Takens F.
Hyperbolicity and Sensitive Dynamics at Homoclinic Bifurcations, Fractal Dimensions and Infinitely Many Attractors, Cambridge University Press, Cambridge, 1993

Peitgen H.-O., Jürgens H. and Saupe D.
Fractals for the Classroom, Part II, Springer Verlag, 1992

Peitgen H.-O. and Richter P. H.
The Beauty of Fractals, Springer Verlag, 1986

Preston C.
Iterates of Maps on an Interval, Springer Verlag, 1983

Ruelle D.
Thermodynamic Formalism, Addison-Wesley, Reading, Mass., 1978

Schuster H. G.
Deterministic Chaos, VCH, Weinheim, 1988

Silverman J.
The Arithmetic of Dynamical Systems, Springer, 2007

Spiegel M. R.
Calculus of Finite Differences and Difference Equations, Schaum's Outline Series, McGraw-Hill, 1971

Steeb W.-H.
A Handbook of Terms Used in Chaos and Quantum Chaos, BI Wissenschaftsverlag, Mannheim, 1991

Steeb W.-H.
The Nonlinear Workbook, sixth edition, World Scientific Publishing, Singapore, 2015

Steeb W.-H.
Problems and Solutions in Theoretical and Mathematical Physics, Third Edition, Volume I: Introductory Level
World Scientific Publishing, Singapore, 2009

Steeb W.-H.
Problems and Solutions in Theoretical and Mathematical Physics, Third Edition, Volume II: Adanced Level
World Scientific Publishing, Singapore, 2009

Steeb W.-H.
Problems and Solutions in Scientific Computing with C++ and Java Simulations, World Scientific Publishing, Singapore, 2004

Strogatz S.
Nonlinear Dynamics and Chaos with Applications to Physics, Biology, Chemistry and Engineering, Addison-Wesley, 1994

van Wyk A. and Steeb W.-H.
Chaos in Electronics, Kluwer, 1997

Walters P.
An Introduction to Ergodic Theory, Graduate Texts in Mathematics 79, Springer, New York, 1982

Papers

Adachi S., Toda M. and Ikeda K., "Recovery of Liouville dynamics in quantum mechanically suppressed chaotic behaviour", J. Phys. A: Math. Gen. **22**, 3291-3306 (1989)

Aizawa Y., "Symbolic Dynamics Approach to the Two-D Chaos in Area-Preserving Maps", Prog. Theor. Phys. **71**, 1419-1421 (1984)

Aizawa Y. and Murakami C., "Generalization of Baker's Transformation", Prog. Theor. Phys. **69**, 1416-1426 (1983)

Aizawa Y., Murakami C. and Kohyama T., "Statistical Mechanics of of Intermittent Chaos", Prog. Theor. Phys. Supp. **79**, 96-124 (1984)

Andrea S. A., "On homeomorphism of the plane, and their embeddings in flows", Bull. AMS **71**, 381-383 (1965)

Anosov D. V., "Ergodic Properties of Geodesic Flows on Closed Riemannian Manifolds of Negative Curvature", Sov. Math. Dokl. **4**, 1153-1156 (1963)

Antoniou I. and Tasaki S., "Spectral decomposition of the Renyi map", J. Phys. A: Math. Gen. **26**, 73-94 (1993)

Aubrey S., "The Devil's Staircase Transformation in Incommensurate Lattices", Lect. Notes in Math. **925**, 221-245 (1982)

Aubrey S., "The Twist Map, the extended Frenkel-Kontorova Model and the Devil's Staircase", Physica **7D**, 240-258 (1983)

Bandt M., Barnsley M., Hegland M. and Vince A., "Conjugacies provided by fractal transformationss I: Conjugate Measures, Hilbert Spaces, Orthogonal Expansions and Flows, on Self-Referential Spaces", arXiv:1409.3309v1

Banks J. and Dragan V., "Smale's Horseshoe Map via Ternary Numbers", SIAM Review, **36**, 265-271 (1994)

Barlow M. T. and Taylor S. J., "Fractional dimension of sets in discrete spaces", J. Phys. A: Math. Gen. **22**, 2621-2626 (1989)

Barnsley M. F. and Demko S., "Iterated function system and the global construction of fractals", Proc. R. Soc. Lond. **A399**, 243-275 (1985)

Beddington J. R., Free C. A. and Lawton J. H., "Dynamic Complexity in Predator-Prey Models Framed in Difference Equations", Nature **255**, 58-60 (1975)

Birkhoff G. D., "Surface Transformations and Their Dynamical Applications", Acta Math. **43**, 1-119 (1920)

Bowen R., "Topological entropy for noncompact sets", Trans. Amer. Math. Soc. **184**, 125-138 (1973)

Bullett S., "Invariant Circles for the Piecewise-Linear Standard Map", Comm. Math. Phys. **107**, 241-262 (1986)

Burton R. and Easton R. W., "Ergodicity of Linked Twist Maps", Springer Lecture Notes in Mathematics, **819** 35-49 (1980)

Chirikov B. V., "A universal instability of many-dimensional oscillator systems", Phys. Rep. **52** 265-379 (1979)

Chirikov B. V., Lieberman M. A., Shepelyansky D. L. and Vivaldi F. M., "A Theory of Modulational Diffusion", Physica D **14**, 289-304 (1984)

Collet P., Eckmann J.-P. and Koch H., "On Universality for Area-Preserving Maps of the Plane", Physica D **3**, 457-467 (1981)

Coppel W. A., "An Interesting Cantor Set", Amer. Math. Monthly **90**, 456-460 (1983)

Crawford J. D. and Cary J. R., "Decay of Correlations in a Chaotic Measure-Preserving Transformation", Physica D **6**, 223-232 (1983)

Curry J. H., "On the Henon transformation", Commun. Math. Phys. **68**, 129-140 (1979)

Daido H., "Onset of Intermittency from Torus", Prog. Theor. Phys. **71**, 402-405 (1984)

Das N. and Dutta T. K., "Hopf Bifurcations on a Nonlinear Chaotic Discrete Model", Int. J. Sci. Eng. Appl. **1**, 127-134 (2012)

Dana I. and Fishman S., "Diffusion in the Standard Map", Physica D **17**, 63-74 (1985)

Derrida B., Gervois A. and Pomeau Y., "Universal metric properties of bifurcations of endomorphisms", J. Phys. A: Math. Gen. **12** 269-296 (1979)

Devaney R., "Reversible Diffeomorphisms and Flows", Trans. Am. Math. Soc. **218**, 89-113 (1976)

Devaney R., "A Piecewise Linear Model for the Zones of Instabilty of an Area-Preserving Map", Physica D **10**, 367-393 (1984)

Devaney R. and Nitecki Z., "Shift Automorphisms in the Hénon Mapping", Comm. in Math. Phys. **67**, 137-146 (1979)

Dragt A. J. and Finn J. M., "Lie series and invariant functions for analytic symplectic maps", J. Math. Phys. **17**, 2215-2227 (1976)

Ebeling W. and Nicolis G., "Word frequency and entropy of symbolic sequences: a dynamical perspective", Chaos, Solitons and Fractals, **2**, 635-650 (1992)

Eckmann J.-P., "Roads to turbulence in dissipative dynamical systems", Rev. Mod. Phys. **53**, 643-670 (1981)

Eckmann J.-P., Koch H. and Wittwer P., "A Computer-Assisted Proof of Universality for Area-Preserving Maps", Mem. Am. Math. Soc. **47:289**, 1-122 (1984)

Falconer K. J., "A subadditive thermodynamic formalism for mixing repellers", J. Phys. A: Math. Gen. **21**, L737-L742 (1988)

Feigenbaum M. J., "Quantitative Universality for a Class of Non-Linear Transformations", J. Stat. Phys. **19** 25-52 (1978)

Feigenbaum M. J., "The universal metric properties of nonlinear transformations", J. Stat. Phys. **21** 669-706 (1979)

Fermi E., Pasta J. and Ulam S., "Studies of Nonlinear Problems. I", Lect. App. Math. **15**, 143-156 (1974)

Feudel U., Pikovsky A. and Politi A., "Renormalization of correlations and spectra of a strange non-chaotic attractor", J. Phys. A: Math. Gen. **29**, 5297-5311 (1996)

Fournier-Prunaret D. and Lopez-Ruiz R., "Basin bifurcations in a two-dimensional logistic map", arXiv:nlin/0304059v1

Froeschle C., "Numerical Study of a Four Dimensional Mapping, Astron. and Astrophs. **16**, 172-189 (1972)

Fujisaka H., Kamifukumoto H. and Inoue M., "Intermittency Associated with the Breakdown of the Chaos Symmetry", Prog. Theor. Phys. **69**, 333-337 (1983)

Fujisaka H., "Theory of Diffusion and Intermittency in Chaotic Systems", Prog. Theor. Phys. **71**, 513-523 (1984)

Fujiksaka H., Ishii H., Inoue M. and Yamad T., "Intermittency Caused by Chaotic Modulation. II", Prog. Theor. Phys. **76**, 1198-1209 (1986)

Fukuda W. and Katsura S., "Solvable Models Showing Chaotic Behavior II", Physica **136**A, 588-599 (1986)

Gefen Y., Aharony A., Shapir Y. and Mandelbrot B. B., "Phase transitions on fractals: II. Sierpinski gaskets", J. Phys. A: Math. Gen. **17**, 435-444 (1984)

Goroff D. L., "Hyperbolic Sets for Twist Maps", Ergod. Th. Dyn. Sys. **5**, 337-339 (1985)

Grassberger P. and Procaccia I., "Measuring the Strangeness of Strange Attractors", Physica D **9**, 189-208 (1983)

Grebogi C., Ott E. and Yorke J. A., "Fractal basin boundaries, long-lived chaotic and unstable-unstable pair bifurcations", Phys. Rev. Lett. **50**, 935-938 (1983)

Greene J. M., "Two-Dimensional Area Preserving Mappings", J. Math. Phys. **9**, 760-768 (1968)

Greene J. M., "A method for determining a stochastic transition", J. Math. Phys. **20**, 1183-1201 (1979)

Greene J. M., MacKay R. S., Vivaldi F. and Feigenbaum M. J., "Universal Behaviour in Families of Area-Preserving Maps", Physica **3D**, 468-486 (1981)

Greene J. M. and Percival I. C., "Hamiltonian Maps in the Complex Plane", Physica D **3**, 530-548 (1981)

Grossmann S. and Thomae S., "Invariant distributions and stationary correlation functions of one-dimensional discrete processes", Z. Naturf. **32a**, 1353-1363 (1977)

Gunaratne G. H., Jensen M. H. and Procaccia I., "Universal strange attractors on wrinkled tori", Nonlinearity **1**, 157-180 (1988)

Hammel S., Yorke J. A. and Grebogi C., "Do Numerical Orbits of Chaotic Dynamical Processes Represent True Orbits?" J. Complexity **3**, 136-145 (1987)

Hardy Y. and Sabatta D., "Encoding, symbolic dynamics, cryptography and C++ implementations", Phys. Lett. A **366**, 575-584 (2007)

Hemmer P. C., "The exact invariant density for a cusp-shaped return map", J. Phys. A: Math. Gen. **17**, L247-L249 (1984)

Hénon M., "Numerical study of quadratic area preserving mappings", Q. Appl. Math. **27**, 291-312 (1969)

Hénon M., "A two-dimensional mapping with a strange attractor", Commun. Math. Phys. **50**, 69-77 (1976)

Harayama T. and Aizawa Y., "Poincaré-Birkhoff Chains in the Critical Regime between Chaos and Torus", Prog. Theor. Phys. **84**, 23-27 (1990)

Hardy Y., Steeb W.-H. and Villet C. M., "Stabilization of Chaotic Systems with Phase Coupling", Z. Naturforsch. **55a**, 847-850 (2000)

Hardy Y. and Sabatta D., "Encoding, symbolic dynamics, cryptography and C++ implementations", Phys. Lett. A **366**, 575-584 (2007)

Hemmer P. C., "The exact invariant density for a cusp-shaped return map", J. Phys. A: Math. Gen. **17**, L247-L249 (1984)

Holmes P. J. and Whitley D. C., "Bifurcations of one- and two-dimensional maps", Phil. Trans. R. Soc. Lond. **A311**, 43-102 (1984)

Horita T., Hata H., Mori H. and Tomita K., "Dynamics on Critical Tori at the Onset of Chaos and Critical KAM Tori", Prog. Theor. Phys. **81**, 1073-1078 (1989)

Hutchinson J. E., "Fractals and Self Similarity", Indiana University Mathematics Journal **30**, 713-747 (1981)

Kaizoji T., "Non-Stationary Chaos", arXiv:1007.3638v1

Kaneko K., "Transition from Torus to Chaos Accompanied by Frequency Lockings with Symmetry Breaking", Prog. Theor. Phys. **69**, 1427-1442 (1983)

Kaneko K., "Fates of Three-Torus. I", Prog. Theor. Phys. **71**, 282-294 (1984)

Kaneko K., "Spatiotemporal Chaos on One- and Two-Dimensional Coupled Map Lattices", Physica D **37**, 60-82 (1989)

Kaneko K., "Fractalization of Torus", Prog. Theor. Phys. **71**, 1112-1114 (1984)

Kaneko K., "Partition complexity in a network of chaotic elements", J. Phys. A: Math. Gen. **24**, 2107-2119 (1991)

Kaneko K. and Bagley R. J., "Arnold Diffusion, Ergodicity and Intermittency in a Coupled Standard Mapping", Phys. Lett. A **110**, 435-440 (1985)

Karney C. F. F., "Long-Time Corrleations in the Stochastic Regime", Physica 8D, 360-380 (1983)

Katok A., "Periodic and Quasi-Periodic Orbits for Twist Maps", Springer Lecture Notes in Physics", **179**, 47-65 (1983)

Katok A., "Bernoulli Diffeomorphisms on Surface", Ann. Math. **110**, 529-547 (1979)

Katsura S. and Fukuda W., "Exactly Solvable Models Showing Chaotic Behavior", Physica **130** A, 597-605 (1985)

Kiyoshi Sogo, "Inverse problem in chaotic map theory", Chaos, Solitons and Fractals, **41**, 1817-1822 (2009)

Konishi T. and Kaneko K., "Diffusion in Hamiltonian chaos and its size dependence", J. Phys. A: Math. Gen. **23**, L715-L720 (1990)

Li W. and Bak P., "Fractal Dimension of Cantori", Phys. Rev. Lett. **57**, 655-658 (1986)

Li T. Y. and Yorke J. A., "Period three implies chaos", Amer. Math. Monthly **82**, 983-992 (1975)

Lindsey K. A., "Shapes of Polynomial Julia Sets", arXiv:1209.0143v2

Lopez-Ruiz R. and Fournier-Prunaret D., "Complex pattern on the plane: Different types of basin fractalization in a 2-D mapping", Int. J. Bifurcation Chaos **13**, 287 (2003)

Lorenz E. N., "Deterministic nonperiodic flows", J. Atmos. Sci. **20**, 130-141 (1963)

Lorenz E. N., "The local structure of a chaotic attractor in four dimensions", Physica **13D**, 90-104 (1984)

Loskutov A. Yu., Rybalko S. D., Feudel U. and Kurths J., "Suprression of chaos by cyclic parametric excitation in two-dimensional maps", J. Phys. A: Math. Gen. **29**, 5759-5771 (1996)

Maeda S., "The Similarity Method for Difference Equations", IMA Journal of Applied Mathematics, **38**, 129-134 (1987)

McDonald S. W., Grebogi C., Ott E. and Yorke J. A., "Fractal basin boundaries", Physica **16D**, 125-153 (1985)

MacKay R. S., "A Renormalization Approach to Invariant Circles in Area-Preserving Maps", Physica D **7**, 283-300 (1983)

MacKay R. S. and Tresser C., "Transitions to topological chaos for circle maps", Physica **19D**, 206-237 (1986)

Martin J. C. and Mora L., "C^2-Perturbations of Hopf's bifurcation points and homoclinic tangencies", Proc. Amer. Math. Soc. **128**, 1241-1245 (1999)

Mather J. N., "Non-existence of invariant circles", Ergod. Th. and Dyn. Sys. **4**, 301-309 (1984)

May R. M., "Simple mathematical models with very complicated dynamics", Nature **261**, 459-467 (1976??)

May R. M., "Biological populations obeying difference equations: Stable points, stable cycles and chaos", Biol. Cybernet. **42**, 221-229 (1982)

McMullen C., "The Hausdorff Dimension of General Sierpinski Carpets", Nagoya Math. J. **96**, 1-9 (1984)

Meiss J. D. and Ott E., "Markov-Tree Model Transport in Area Preserving Maps", Physica D **20**, 387-402 (1986)

Mori H., So B.-C. and Ose T., "Time-Correlation Functions of One-Dimensional Transformations", Prog. Theor. Phys. **66**, 1266-1283 (1981)

Mori H., Okamoto H. and Ogasawara M., "Self-Similar Cascades of Band Splittings of Linear Mod 1 Maps", Prog. Theor. Phys. **71**, 499-512 (1984)

Morse M. and Hedlund G. A., "Symbolic dynamics I", Am. J. Math. **60**, 815-866 (1938)

Moser J. K., "The Analytic Invariants of an Area Preserving near a Hyperbolic Fixed Point", Comm. Pure and Appl. Math. **9**, 673-692 (1956)

Moser J. K., "On the Integrability of Area-preserving Cremona Mappings near an elliptic fixed point", Bol. Soc. Mat. Mexicana, 176-180 (1961)

Moser J. K., "On invariant curves of area-preserving mappings of an annulus", Nachr. Akad. Wiss. Göttingen Math. Phys. Kl. **2**, 1-20 (1962)

Nakamura K. and Hamada M., "Asymptotic expansion of homoclinic structures in a symplectic mapping", J. Phys. A: Math. Gen. **29**, 7315-7327 (1996)

Newhouse S., Palis J. and Takens F., "Bifurcation and Stability of Families of Diffeomorphisms", Publ. Math. I.H.E.S. **57**, 5-72 (1983)

Nguyen H. D., "A Digital Binomial Theorem", arXiv:1412.3181v1

Oseledec V. I., "A Multiplicative Ergodic Theorem: Lyapunov Characteristic Numbers for Dynamical Systems", Trans. Moscow Math. Soc. **19**, 197-231 (1968)

Ott E., "Strange attractors and chaotic motions of dynamical systems", Rev. Mod. Phys. **53**, 655-672 (1981)

Ott E., Withers W. D. and Yorke J. A., "Is the Dimension of Chaotic Attractors Invariant under Coordinate Changes?", Journal of Statistical Physics, **36**, 687-697 (1984)

Palmore J. I. and McCauley J. L., "Shadowing by Chaotic Orbits", Phys. Lett. A **122**, 399-402 (1987)

Percival I. C., "Variational Principles for Invariant Tori and Cantori", AIP Conference Proceeding, **57**, 302-310 (1979)

Percival I. C., "Chaotic Boundary of a Hamiltonian Map", Physica D **6**, 67-77 (1982)

Pearse E. P. J., "Canonical Self-Affine Tiling by Iterated Functionen Systems", arXiv:math/0606111v2

Pesin Ja. B., "Ljapunov Characteristic Exponents and Ergodic Properties of Smooth Dynamical Systems with an Invariant Measure", Soviet Mathematics-Doklady **17**, 196-199 (1976)

Ping P., Xu F. and Wang Z.-J., "Color Image encryption based on Two-Dimensional Cellular Automata", **24**, 1350071-1-14 (2013)

Pomeau Y. and Manneville P., "Intermittent transition to turbulence in dissipative dynamical systems", Commun. Math. Phys. **74**, 189-197 (1980)

Radons G. and Stoop R., "Superpositions of Multifractals: Generators of Phase Transitions in the Generalized Thermodynamic Formalism" J. Stat. Phys. **82**, 1063-1080 (1996)

Rimmer R., "Symmetry and Bifurcation of Fixed Points of Area-Preserving Maps", J. Diff. Eqns. **29**, 329-344 (1978)

Saito N., "Baker's Transformation and Invariant Measure", J. Phys. Soc. Jpn **51**, 374-378 (1982)

Sarkovskii A. N., "Co-Existence of Cycles of a Continuous Mapping of a Line onto itself", Ukranian Math. Zh. **16**, 61-71 (1964)

Schäfer A. and Müller B., "Bounds for the fractal dimension of space", J. Phys. A: Math. Gen. **19**, 3891-3902 (1986)

Schmidt G., "Stochasticity and Fixed Point Transitions", Phys. Rev. A **22**, 2849-2854 (1980)

Schmidt G. and Bialek J., "Fractal Diagrams for Hamiltonian Stochasticity", Physica **5D**, 397-404 (1982)

Schröder E., "Ueber iterirte Functionen", Math. Ann. **3**, 296-322 (1870)

Schwegler H. and Mackey M. C., "A simple model for the approach of entropy to thermodynamic equilibrium", J. Phys. A: Math. Gen. **27**,1939-1951 (1994)

Sinai Ya. G., "The Construction of Markov Partitions", Funct. Ana. Appl. **2**, 70-80 (1968)

Sinai Ya. G., "Gibbs Measures in Ergodic Theory", Russ. Math. Surv. **27**, 21-69 (1972)

Steeb W.-H. and van Wyk M. A., "Invariants and Chaotic Maps", Int. J. Theor. Phys. **35**, 1253-1257 (1996)

Steeb W.-H., Solms F., Tan Kiat Shi and Stoop R., "Cubic Map, Complexity and Ljapunov Exponent", Physica Scripta **55**, 520-522 (1997)

Stefan P., "A theorem of Sharkovsky on the existence of periodic orbits of continuous endomorphisms of the real line", Comm. Math. Phys. **54**, 237-248 (1977)

Takahashi Y. and Oono Y., "Towards the Statistical Mechanics of Chaos", Prog. Theor. Phys. **71**, 851-853 (1984)

Theiler J., Mayer-Kress G. and Kadtke J., "Chaotic attractors of a locally conservative hyperbolic map with overlap", Physica D **48**, 425-444 (1991)

Tsuda I., "On the Abnormality of Period Doubling Bifurcations", Prog. Theor. Phys. **66**, 1985-1994 (1981)

Umberger D. K. and Farmer J. D., "Fat Fractals on the Energy Surface", Phys. Rev. Lett. **55**, 661-664 (1985)

Widom M. and Kadanoff L. P., "Renormalization Group Analysis of Bifurcations in Area Preserving Maps", Physica D **5**, 287-292 (1982)

Wojtkowski M., "A Model Problem with the Coexistence of Stochastic and Integrable Behaviour", Comm. in Math. Phys. **80**, 453-464 (1981)

van Wyk M. and Steeb W.-H., "Stochastic Analysis of the ZigZag Map", Int. J. Mod. Phys. C **14**, 397-403 (2003)

Yalcinkaya T. and Lai Y.-C., "Chaotic Scattering", Computers in Physics, **9**, 511-539 (1995)

Yamaguchi Y., "Renormalization Group Approach to Universal Metric Properties of Whisker Mapping", Prog. Theor. Phys. **72**, 694-709 (1984)

Yoshida T. and Tomita K. "Characteristic Structures of Power Spectra in Periodic Chaos", Prog. Theor. Phys. **76**, 752-767 (1986)

Young L. S., "Dimension, entropy and Lyapunov exponents, Ergodic Theory Dynamical Systems **2**, 109-124 (1982)

Young L. S., "Entropy, Lyapunov Exponents and Hausdorff Dimension in Differentiable Dynamical Systems", IEEE Transaction Circuits and Systems, CAS-30, 599-607 (1983)

Zisook A. B. and Shenker S. J., "Renormalization Group for Intermittency in Area-Preserving Mappings", Phys. Rev. A **25**, 2824-2826 (1982)

Index

Printed in the United States
By Bookmasters